RUSSIAN AND EAST EUROPEAN STUDIES

The Education of a Russian Statesman

RUSSIAN AND EAST EUROPEAN STUDIES

THE EDUCATION OF

A RUSSIAN

STATESMAN

The Memoirs of Nicholas Karlovich Giers

EDITED BY

CHARLES AND BARBARA JELAVICH

University of California Press   1962

BERKELEY AND LOS ANGELES

University of California Press
Berkeley and Los Angeles
California

Cambridge University Press
London, England

© 1962 by The Regents of the University of California
Library of Congress Catalog Card Number: 62-14297

Printed in the United States of America

For Robert Joseph Kerner

PREFACE

ONE OF THE DIFFICULTIES in the study of Russian internal and foreign policy in the nineteenth century is the insufficiency of material on the lives and thought of the principal statesmen and administrators. This deficiency is particularly apparent for the careers of the three men chiefly responsible for the conduct of Russian foreign relations—K. V. Nesselrode, A. M. Gorchakov, and N. K. Giers. Although state papers presumably drafted by them, and a collection of Nesselrode's letters,[1] exist, there is little other material available from which one can draw a clear picture of their personal characteristics and their views in general. The following memoirs, therefore, are in a sense unique in that they provide a close and intimate glimpse into the mode of life and development of one of Russia's leading ministers. The document has also the advantage that it was written only for the immediate family of Giers and was not planned for publication. The questions discussed are therefore those which the author intended for the information and interest of his children, and he is consistently open and frank in his expression of opinion and in his judgment of others.

The original manuscript of these memoirs is now in the possession of Serge Giers, a grandson of N. K. Giers, who lives in France. A microfilm copy is available in the General Library of the University of California, Berkeley. The work was not only left uncompleted, but there is no indication that Giers himself reread the draft in order to make corrections or additions. The style is therefore that of a man who writes for his own pleasure and in order to chronicle his activities for his relatives. His sentences are generally short and the structure is at times awkward. He tends to repeat some words, often within the same sentence, and he is especially attached to the frequent use of "very," "completely," "extremely," "all," "positively," and the phrase "as I have already mentioned." In order to provide a smoother translation the editors have joined or recast short sentences and have omitted words or short phrases where such deletions did not change the meaning of the sentence. An attempt has, nevertheless, been made to reproduce in English the general style of the author, who was a direct and simple, rather than a polished, writer.

In the editing and translating of the following pages difficulties arose in connection with the spelling of proper names in Russian and Rumanian. Although exceptions have been made, the following rules have been observed in the translation: Names of people and places have been rendered in the spelling of their national origin except where a standardized form is used in most English books. It is, for example, Giers, not Girs, and Jassy, not Iași. All proper names in Russian have been transliterated according to the Library of Congress system, which has also been used for those persons in Russian society or service where the name is obviously of foreign origin, for instance, Engel'gardt (Engelhardt), and Vitgenstein (Wittgenstein). An exception has been made for Egor Matveevich Cantacuzino (Kantakuzin) because he appears more often in the Rumanian sections where another system of spelling was used. Here Rumanian names which in the original are given in the French form and, of course, are written in the Cyrillic alphabet, have been changed into Rumanian. Place names throughout are given according to the political status of the territory at the time of the events recorded in the memoir. Thus Bessarabian cities are in Russian because the land was then in Russian possession. Complete consistency in these matters was, however, impossible, especially in connection with places situated in and around St. Petersburg. Here the editors have attempted to employ the forms which they believe are most familiar to the reader of Russian history and literature. Thus, for instance, it is Nevskii Prospect, not Neva Boulevard; Tsarskoe Selo, not Tsar's Village; Winter Palace, not Zimnyi Dvorets. When a place is relatively less known, the Russian spelling has been used, as in Kamennyi Island or Sinyi Bridge. This decision has of necessity been arbitrary.

The translation was done by the editors and by Mrs. Nuncia Lodge and Dr. Stephen Lukashevich. Mrs. Lodge prepared a draft of the translation and Dr. Lukashevich checked the entire manuscript. The editors wish to express their great appreciation to Professor Stephen Fisher-Galati of Wayne University and Dr. Emanuel

Turcyzinski of the University of Munich, who read the sections on Rumania and offered valuable suggestions and criticisms. Mr. Frederick Kellogg aided greatly in the location of Rumanian cities and the identification of individuals. Mr. Stephen Soudakoff gave welcome advice in connection with problems which arose in association with the section on Russia. The map is the product of the expert drafting of Professor Norman J. G. Pounds of Indiana University and Mr. Robert C. Kingsbury. Mr. Max E. Knight of the Editorial Department, University of California Press, gave excellent editorial criticism. We wish also to express our thanks to the Center for Slavic and East European Studies of the University of California, Berkeley, for aid in the preparation of this book. A particular word of appreciation must again be given to Mr. Serge Giers, who in connection with this manuscript, as with our previous publications that used his private papers,[2] offered every possible assistance and showed real appreciation for the needs of scholarship.

C. AND B. J.

1962

CONTENTS

Childhood and Education

(1820–1841)

EDITORS' INTRODUCTION TO PART ONE

THE FOLLOWING PAGES present the unfinished memoirs of Nicholas Karlovich Giers, the Russian foreign minister from 1882 to 1895. The work is divided into two parts, the first covering the author's childhood, his education in St. Petersburg, and his first appointment in the Asiatic Department of the Foreign Ministry, the second his experiences in Moldavia where he was sent in 1841 as second dragoman at the consulate in Jassy. These memoirs were written while Giers was minister in Stockholm from 1873 to 1875. When he was called back to St. Petersburg in 1875 to assume the more responsible position of assistant foreign minister and head of the Asiatic Department, he was unable to find time to complete this work.

The memoirs are published in their entirety with the exception of the first pages which contain a detailed chronology of the Giers family and the fate of its various members. Here Giers traces his ancestry back to a companion of Gustavus Adolphus, a colonel in the artillery who died in the Thirty Years' War. The first direct relative whom Giers can name, however, was a certain Lorenz Giers who likewise served in the Swedish army. One of his sons, also named Lorenz, was with Charles XII and was made prisoner at the battle of Poltava. His son, the third in the direct line with the name of Lorenz, was born in 1718 and migrated to Russia about 1750. His son, Karl Lavrent'evich Giers, served under the Polish king Stanislaus Poniatowski, then returned to Russia where he joined the Ministry of Finance, and became the director of the customs of the district of Grodno. He married a German, Sofia Ulf, and had five children. His second son, Karl Karlovich, the father of Nicholas Karlovich, in 1810 married Anna Petrovna Litke (Lütke) whose ancestors had come to Russia from Germany during the reign of Empress Elizabeth and who distinguished themselves in the Russian administration.

Giers's background was thus typical of that of the many families who moved from central Europe and the Baltic area to Russia in the eighteenth century to enter Russian state service. In fact, at that time German names came to dominate in the upper echelons of the Russian army. The diplomatic service was similarly filled by those of foreign extraction. In the first part of the nineteenth century when a Pole, Czartoryski, was succeeded as foreign minister by the Greek Capodistrias, who in turn was supplanted by the German Nesselrode, a non-Russian background was no hindrance to advancement in Russian state service. The first significant change in this practice occurred after the Crimean War and an even stronger reaction occurred during the reign of Alexander III.

Giers's ancestry thus had a major significance in his career. He always had to meet the objections that he was a "German" and a Lutheran. It is interesting to note that throughout his memoirs Giers affirms his poor knowledge of German [1] and his general lack of sympathy for the German elements in his surroundings. He emphasizes his Swedish as against his German ancestry despite the fact that his Swedish ancestors appear usually to have married German wives. In the same manner Giers refers to his sympathy with the Orthodox Church and its rites although he never abandoned the Lutheran faith.

Despite his family relationship with those who held major positions in Russian administration, there was little in Giers's early career which foreshadowed his eventual advancement to one of the highest state offices. As related in these memoirs, chance circumstance together with the author's diligence and seriousness as a student enabled him to enjoy the best education tsarist Russia had to offer—that of the Lyceum of Tsarskoe Selo. Thereafter for many years his career was not distinguished. He describes the events of his first disappointing years in the Asiatic Department and his subsequent long residence in Moldavia. Although he was personally disheartened by the slowness of his advance, the years in the Danubian Principalities, the name by which Moldavia and Wallachia were known, resulted in an event which had a decisive influence on his future. In 1849 he married Olga Cantacuzino, whose mother was a sister of A. M. Gorchakov, the Russian foreign minister from 1856 to 1882. Olga corresponded frequently with her uncle and Gorchakov did what he could to assist his niece and her family.

Certainly, before Gorchakov's assumption of the ministry, Giers did not advance quickly. In 1849 he was the diplomatic agent attached to General Lüders during the Russian occupation of Hungary. Thereafter he became first secretary of the embassy in Constantinople, and in 1852 he returned to Jassy as consul. In the Crimean War he was head of the diplomatic chancellery of Count Stroganov at Odessa. After a period in Alexandria as consul general, he was sent as consul general to Bucharest for the years 1858 to 1863. His first major appointment came when he was designated minister to Persia, where he remained from 1863 to 1868. After that he received the very agreeable position of minister in Berne, and in 1872 he was sent to Stockholm. In 1875 he was advanced to the position of Gorchakov's deputy, and after 1878, because of the increasing senility of his superior, he functioned as foreign minister although he did not officially hold the post until 1882.

As foreign minister Giers never had the independence of action or the direct responsibility enjoyed by those with similar positions in other countries. Unlike, in particular, his great contemporary Bismarck, Giers was unable to formulate and implement policies of his own design. It was rather the tsars under whom he served, Alexander II and Alexander III, who directed foreign affairs and made the decisions after consultations with their ministers. Giers thus functioned more as a secretary for foreign affairs than a minister. Nevertheless, he was able to make his influence felt. A constant advocate of compromise and moderation, a foe of adventures and adventurers, Giers sought always to preserve Russia from foreign conflicts. Like his predecessor Gorchakov, Giers supported the arguments of those who believed that Russia should avoid foreign entanglements—above all, any that might lead to war—and instead should concentrate on internal reform. He was thus a convinced proponent of the Three Emperors' Alliance (Dreikaiserbund) of Germany, Russia, and Austria-Hungary because he believed that it offered the best hope of security and peace to Russia. In the 1890's he accepted the policy of alignment with France for the same objectives. As the advocate of the moderate course and an honest and straightforward policy, Giers was able to offer guidance and a restraining hand to the tsars he served.

Giers's position under Alexander II and Alexander III, which was illustrative of the normal relationship of tsar and minister from the time of Nesselrode,[2] demanded certain attributes in the character of those who ran the Foreign Ministry. The memoirs show clearly why Giers filled so successfully the role which his office demanded. The reader is impressed by the fact that young Giers likes everyone. The only person who is criticized with severity is his first superior, L. G. Seniavin. In his later years the ability of Giers to work well with others and to win their trust and approval made him a most valuable official. The letters written to him during his ministry reflect the confidence and even affection which he was able to win from his colleagues. His avoidance of intrigue and backstage maneuvering was in sharp contrast to the activities of other Russian diplomats such as Saburov, Ignat'ev, and Mohrenheim.

The industriousness which Giers showed as foreign minister can also be seen in these pages in the habits of the conscientious schoolboy who worked so diligently and studied so hard. It is further reflected in his dislike of school "pranks" and in his obvious popularity with the school authorities who approved of his seriousness and decorum. Without a personal fortune and with a large family to support,[3] Giers throughout his career had to rely on his own efforts for advancement. Those who had close connections with the great Russian families and great fortunes could afford to play, but Giers had only his government allotment. The combination of natural diligence and necessity reinforced his tendencies toward conservatism and moderation. He could not indulge in political adventures or adopt methods which would alienate his superiors.

Giers as a student was obviously delighted with the life of the Russian court and was happy to be included in its functions in any capacity. Throughout his career he accepted Russian autocracy as such, and there is no evidence that his mind ever turned to questions of politics of a general nature. Like his fellow ministers M. K. Reutern and D. A. Miliutin, he realized the necessity for reform, but only in the details of administration which would allow the government to function more effectively. In conviction and character Giers was thus well suited to his position as servant and secretary to the tsar.

Despite the fact that the temperament of Giers throughout his career and as reflected in these pages was moderate and tolerant in most matters, his remarks in these memoirs show also a strong aversion to the Jews, which, particularly in view of the tragedy of our own day, call for special comment. Although Giers judges negatively the Germans, the Poles, and most strongly the Rumanian boyars and the Orthodox Church in the principalities, he is particularly disdainful of the Jewish people he encounters and to whom he always refers by the derogatory term *Zhid* rather than the more correct *Evrei*. Unfortunately, this attitude reflected the outlook of a great number of his generation and was expressed in the discriminatory measures taken against the Jews after 1880 and in the pogroms which were carried out with at least official indifference under both Alexander III and Nicholas II. Anti-Semitism in tsarist Russia was part of the same general policy which led to the gradual repression and restriction of the local privileges and rights of the Poles, Finns, and Baltic Germans, and the religious sects such as the Dukhobors and the Stundists.

In the beginning of the memoirs following the detailed account of the Giers family and its arrival in Russia, the author tells how his mother met and married his father, who held the modest position of frontier postmaster at Radzivilov. His mother's sister, Natalia Petrovna Litke, who came to Radzivilov to attend the wedding, met there Admiral Ivan Savvich Sul'menev, who was returning to Russia from service in the Mediterranean. A double wedding was subsequently decided upon and thereafter the Sul'menevs left for St. Petersburg while the parents of Giers remained in Radzivilov. Giers then continues his description of his childhood in this region.

I

MY PARENTS' HOME (1820–1829)

IN THE 1820'S THE VILLAGE of Radzivilov, which did not even merit the designation of town, played a significant role both commercially and politically. Situated right on the border, opposite the Austrian town of Brody, Radzivilov was the most important frontier point on the western border of Russia. Until the introduction of steamship communication, which began only in the 'thirties of this century, most of our connections with Europe were through Radzivilov. This place was particularly important for our relations with Vienna which in those days was regarded as the most important political center of Europe. The events of tremendous importance which were taking place in the West in the early years of the present century, and which were destined to have so decisive an influence on Russia, inspired our government to have a reliable and gifted man in Radzivilov; he was to facilitate our relations with the Austrian government with which we tried to preserve friendly relations despite the influence of Napoleon which prevailed there at the time. In any case, it was important to watch what was going on at this frontier point of the empire. The choice fell upon my father, and he was sent to Radzivilov as border postmaster. This position gave him the opportunity to carry out successfully the mission entrusted to him, as is testified by the correspondence left upon his death (now in the hands of my older brother, Alexander Karlovich) and, particularly, by the letters from the head of the Ministry of Foreign Affairs, State Secretary Count Nesselrode. In 1812 my father rendered important services when Austria was forced by Napoleon to send an army against us led by Prince Schwarzenberg. At that time General Tormazov's corps was moving against Austria in Volynia.[1] Officially, both enemy armies avoided conflicts, awaiting the time when they could cast off pretenses and could fraternize. This moment came soon after Napoleon was forced out of Russia. My father was witness to this unique campaign and was useful to General Tormazov because of his relations with the Austrian authorities. When events took another direction, the sphere of my father's activity was greatly diminished, but he sought no other field. He liked the life at the border at Radzivilov, where he established friendships with the best Polish families and where he could give full vent to his passion for entertaining. None of the Russian travelers who went abroad or returned from there through Radzivilov (and in those days, as I have remarked, it was one of the chief points of communication with Europe) could pass by this village without visiting my father's home for a few days. This mode of life also satisfied the needs of his family by helping him to bring up and educate his children. Aside from the excellent social life which we thus enjoyed in Radzivilov, we also frequently met famous foreign artists who visited Russia, and, in appreciation for the hospitable reception, they offered us pleasures which could not be found in any other provincial town.

The blessing of the Lord did not leave my parents in their modest circumstances. Conjugal happiness was their worthy reward for their virtues which brought them the love and respect of those who knew them. In spite of their limited means, they were never in need. Thanks to the low cost of living in those days in the region where they resided, they were able to live well and could give their children a good education, and there were ten of us. The oldest was Peter, who died in childhood. He was followed by my sister, Sofia, who was born in 1813 and died in 1836. She was married to Collegiate Counselor [2] Paul Sitnikov, who also died a long time ago. After her came my brother, Alexander, born May 29, 1815. He was followed by my sister Emilia Speranskaia, born in 1817, who died in 1866. After her came Valeria, born August 6, 1818, married to Anferov. On May 9, 1820, the writer of these lines, Nicholas, was born. He was followed by Iulia, born in 1822, married to Colonel Kolonov. After her came my brother Fedor, born January 4, 1824, followed by Konstantin, who died at the age of two, and finally my sister Anna, born in June of 1829 and married to Mikhail Fedorovich Untilov. Since our parents were Lutherans, we were all christened in that faith.

We did not live long under the parental roof. Except for my sisters Sofia and Iulia, we left our home to obtain our education

which we received in various schools in Petersburg. After that each
went his own way. All of us are married and have our own families.
As we shall see, fate has scattered us so that we seldom have occasion to meet, particularly myself, who has been living abroad for
the past thirty years. I must confess to my shame that most of my
nephews and nieces are unknown to me. Nevertheless, they are all
dear to me and they need not doubt my sincere interest in them.

Although I left my home when I was only nine years old, the
recollections of my early childhood are still vividly impressed upon
my mind. It seems to me that if fate should again bring me to
Radzivilov, I would at once recognize not only the house where
we lived, but the streets and main buildings of that small village.
I even remember the location of the various rooms in our house,
so deeply did the early impressions of my childhood engrave themselves into my mind!

As I have already said, Radzivilov was situated at the very border, opposite the Austrian town of Brody, known for its commercial activities. Unfortunately, almost the entire trade is in the hands
of the Jews whose number is large here as well as along the length
of our western frontier. There they took possession also of all the
industry so that one could not get along without them. Because
of this filthy population Radzivilov does not present a very attractive picture. However, this drawback is compensated by the
excellent climate and the presence there of several worthy families,
local Volynia landowners, as well as various administrators, or, at
least, so it was in our time. I recall a magnificent stone house with
a garden which belonged to Count Turno. The stretch of long
avenue extending from it was the main street of Radzivilov. The
back part of our large garden faced this avenue, and we entered it
through a small garden door. In front of our house, facing the
street, was a spacious yard with a well in the center. To the right
were the out-buildings and on the left the post office. In back of
the post office was another house with a yard where my father's
assistant, Il'iashevich, lived with his large family. As far as I recall,
these buildings were the property of Il'iashevich. The long, one-
story house was not distinguished by elegant architecture because
as the family increased my father made additions to it. But it was
very comfortable. Everyone had his corner there, and the recep-

tion rooms were large. What was important was that the gentle and sincere hospitality of the owners attracted to our house not only the fine society of Radzivilov, but also that of the surrounding area. How well do I recall the guests who gathered in our home on family holidays, coming from even as far away as Dubno, which was fifty versts from Radzivilov, and where our Divisional Headquarters was then stationed.

My father was of medium height. The fine features of his face and the kind expression of his eyes have never been effaced from my memory although I never had the fortune of seeing him again after our parting in the ninth year of my life. He was exceptionally kind and lenient. Assisting those in need was a genuine pleasure for him. He felt these Christian sentiments toward all without exception. Even the Jews enjoyed his charity. I recall my mother telling me that a Jewish watchmaker once came to him and asked for work. Since all the clocks in the house were in order, there was no need for the services of this Jew. This is what my father told him, but the man pleaded for an opportunity to earn something to help his family which was in dire need. He did not want to accept alms. Whereupon my father took a brand-new clock from the table and, in spite of the pleading of my mother that the Jew would, of course, only ruin it, he gave it to him to "fix." "How else," he said to my mother, "could I help this poor man!" My father was unusually kind and lenient with his children. A slight reprimand was all we heard from him when any one of us misbehaved. As for strict punishment, he would have nothing to do with it. We must remember that he acted thus in the days when strictness was the rule of education everywhere.

My father found a most worthy companion in my mother. According to the testimony of those who knew her, she was the kindest of the Litke family, all of whom were noted for their kind hearts. With her tender concern for her children she devoted herself entirely to them and took charge of their education up to a certain age. I received my first lessons in Russian, French, and religion from my mother. To this day I recall with what love and patience she taught me and how she rejoiced in my early progress. For my part I was greatly attached to her and, no doubt because of this, had the reputation of being her favorite. I thought, however, that my mother loved all of her children equally, as her conduct toward us proved after we grew up and left the family home.

I recall well how my mother looked because I had the opportunity of seeing her in Petersburg as late as 1835 when I had reached the age of sixteen. She was then past forty, above medium height, and her hair was light brown. Although she was not considered to be a beauty, she nevertheless charmed everyone by her manner, education, fine mind, and frank character. She was of a cheerful disposition and particularly enjoyed giving pleasure to and amusing her children. As a result our childhood in our family home was a very happy one.

The older of our children always had a tutor or a governess. I recall a candidate for a degree in theology, Lippe, who for the most part taught my brother Alexander; also Mlle. Virginie and later Mlle. Etinger who taught my sisters Emilia and Valeria. As far back as I can remember, our sister Sofia was already a grown-up young lady and took lessons in music only.

My recollections about the time I spent in my parents' home lack continuity. Many events stand out clearly in my memory, but they are isolated and disconnected. Thus even now I can see clearly before me my brother Alexander with his tutor Lippe explaining the Bible to him. Yet I was hardly six years old at the time. On the other hand, I definitely do not recall the slightest circumstance connected with Alexander's departure from Radzivilov much later, in 1827. Likewise I remember our children's games with Emilia and Valeria, but not their departure for Petersburg, which followed in 1828.* Valeria, who was always gay, was the life of our games. Emilia was somewhat clumsy and absent-minded, but she was

* I still recall another incident which created excitement in our household. Once my mother went for a walk in our garden with my little sister Iulia who was then about four years old. Suddenly at a turn into one of the paths she heard a shot and felt something hit her foot. She was wounded by a rifle shot. It seems that my father's assistant Il'iashevich had decided to chase a hare in the garden and had shot at it the very second that my mother unexpectedly turned into the path where he stood. Fortunately, the wound from the small shot was trivial, but my mother was more disturbed by the thought of the danger to little Iulia. Had the shot turned slightly in the direction of the child, it would have hit her straight in the chest.

better in her studies than our whirligig Valeria. Once we spent the summer in the beautiful city of Kremenets while our house in Radzivilov was being remodeled. This was, no doubt, in 1827 because my brother Alexander was not with us. That summer was particularly memorable to me because I wore a jacket for the first time and I spilled some milk on it that same day. This misfortune grieved me so much that I cried the whole evening. In the free hours we played in the magnificent garden of the house we occupied, which, as I recall, belonged to one Maievskii. Our games were headed more than supervised by Mlle. Virginie, who was of an unusually cheerful disposition. I recall how she made us act out the fables of La Fontaine.

In those days Kremenets was famous for its Lyceum. It is well known that thanks to Prince Adam Czartoryski,[3] who for some time enjoyed a tremendous influence over Emperor Alexander I, until he openly betrayed him, all possible methods were employed for the Polonization of the western region through educational institutions, where all subjects were taught in the Polish language. The number of these institutions was increased, and they were improved. Russia was to pay heavily because of this course, perfidiously pursued with the secret intention of preparing the ground for the overthrow of the government in that region. The inconceivable blindness of Emperor Alexander bore the bitterest fruits for Russia. All institutions of learning in the western provinces, beginning with the famous Vilno University, became genuine hotbeds for the Polish rebels, and *this* in a region where the bulk of the population unquestionably belonged to the Russian nationality and where the Polish population constituted but one tenth of the total.

If we analyze carefully the actions with regard to the Polish question taken by Emperor Alexander Pavlovich, so tenderly beloved by his people and referred to as the Blessed, we cannot help but regret that because of his insistence at the Congress of Vienna, he was able to form a Polish kingdom from the Duchy of Warsaw and to annex it to Russia with an independent and, in addition, a constitutional government. It is a pity that he refused to follow the advice of the profound statesman, Count Pozzo di Borgo,[4] who with remarkable foresight prophesied the consequences of this measure. Had our tsar at the Congress of Vienna left the Poles in their former position, the one in which they were following the third partition of Poland, no Polish question would have existed, and our western section of the country would have again been merged with Russia! [5]

Be it as it may, but at the time I am describing the Kremenets Lyceum was at its prime, and thanks to this circumstance we were able to have the best instructors in all subjects. There we became acquainted with many Polish families. With one of them (I do not recall the name) we took dancing lessons which were very gay.

While in Kremenets I used to go horseback-riding on a small bay horse that my father gave me. Accompanied by our music teacher, G. Attse, I rode daily around the city whose suburbs are unusually picturesque. I do not know how Attse happened to come to Radzivilov, where he lived for several years and always in our family. His wife, an intelligent, educated, and handsome woman, came from a fine family. It is hard to understand what prompted her to marry so ugly a man as Attse, who, in addition, had a difficult and unpleasant disposition and, moreover, did not possess sufficient musical talent to hope to earn enough money by his profession to provide for his family. However, Attse was an honest and good man, and greatly attached to our family. His wife became a sincere and bosom friend of my mother. Of his two daughters, with whom I spent my childhood years, the elder, Lubin'ka, died in Radzivilov when she was seven or eight years old. And the younger, Mina, married the famous pianist Gerke in Petersburg.

I recall also our memorable trip to Pochaev which was then still a Uniate monastery. This magnificent cloister made a great impression upon me.

By fall we had moved back to Radzivilov where our house had been thoroughly remodeled during our absence. About this time some Sardinian marquis (or at least he called himself that), whose name I cannot remember, came to stay with us as my tutor. He was a native of Savoy and, consequently, had an excellent command of the French language. Forced to leave his country because of his activity in the Carbonari societies and greatly persecuted after the Congress of Verona, he left to find asylum in Russia. On crossing the border into Radzivilov, he came to my father, who, sensing in him the qualities of a decent and educated man, offered him the post of tutor for me. Thus the marquis became my tutor. I am indebted to him, if not for my knowledge in general, at least for the French language. He was cheerful and a great talker, and we chatted a great deal. Often he would tell me about his country, his estates, and the magnificent castle which he lost as a consequence of political circumstances. He assured my parents that if all this were ever returned to him, he would make me his heir. The marquis was already of middle age and not married. He liked me very much and treated me most kindly, which, however, did not prevent him from forgetting me completely when circumstances made it possible for him to return to his native land where, as I learned later from my mother, he did succeed in restoring his former wealth and position.

My recollections of Radzivilov are most pleasant. In addition to our neighbors, the Polish landowners, with whom my parents were on friendly terms, and among whom the family of Count Olizar is particularly memorable to me because of Charles Olizar, a boy of my age who was my best friend in our children's games, we also saw a great deal of some fine Russian families who lived in Dubno. The divisional headquarters was situated there, also various temporary administrative departments, whose purpose I do not know. I recall particularly the families of Generals Bublik and Gogol. The wife of the latter was first married to Count Olizar.* In those days Dubno was the seat of a special commission composed of officers from the general staff set up to delimit the boundaries between Russia and Austria. The head of this commission was the well-known *General-en-Chef* Dovre. Attached to him were Colonels Count Mavriki Gauke and Boianovich. All of them were on the friendliest terms with my father.†

* The elder son of the first marriage of this Gogol made a fine career for himself. He held the post of tutor to the late grand duke Nicholas Alexandrovich. At present he is adjutant general and chief superintendent of the town of Tsarskoe Selo and the palaces there.

† I saw Dovre and Gauke later in Petersburg. Count Gauke died around 1836 with the rank of major general in the retinue of His Imperial Majesty. Tsar Nicholas Pavlovich was particularly kind to him after the glorious death of his brother, also a general, who was killed by the rebels in 1830 in War-

I recall also Major-General Etinger, Colonels Baron Alexander Vrangel' and Ungebauer, the commanders of the chasseurs regiments that were stationed in those regions, also the artillery captain Kutuzov and many others who used to come to see us from Kremenets where the brigade staff was stationed.

We had a large circle of acquaintances in Radzivilov. We used to see most frequently the charming family of Baron Shtomelberg, the chief of the custom's district—the successor, if I am not mistaken, of Anton Ivanovich Litke. In those days, as I have already mentioned, travelers who were returning to Russia frequently stopped with us. Among them were remarkable people, as, for example, Field Marshal Prince Vitgenstein [Wittgenstein],[6] Count Arakcheev,[7] and others. I recall particularly the visit of Arakcheev, who was met and seen off with many honors. I see this dignitary as if it were today, sitting unceremoniously in a bright dressing gown in his traveling coach, while the entire court yard was filled with military and civilian employees in parade uniform who were awaiting his appearance. A convoy of customs mounted patrols followed his carriage.

In reminiscing about my life in my parents' home, I cannot pass

---

saw when he saved the life of Grand Duke Konstantin Pavlovich. His son, our Radzivilov friend, was made a page and served in the Guards. But, in spite of the tsar's favors, he became an inveterate enemy of Russia. In 1862 he went over to the side of the Polish rebels and commanded a strong detachment of revolutionaries under the name of Bossak. After the revolt was suppressed, he escaped abroad and joined the famous Garibaldi who regarded him as his closest collaborator. He bore the title of general, and, in command of a brigade, participated with Garibaldi in the war of 1870. He fell in the battle of Dijon. His sister Iulia, lady in waiting at the court of His Imperial Majesty, married Duke Alexander of Darmstadt, brother of our empress, and was named Duchess of Battenberg.

I heard that the events of 1830 placed Colonel Boianovich in a very serious position. He did not share the opinion of his compatriots with regard to the revolt, but he also refused to fight against them. The moral affliction he endured from this position greatly affected his health. He moved to Cracow where he died soon afterward. He became greatly attached to me, and on my departure from my parents' home for Petersburg made me a present of a book which I preserved for a long time. Its title was *Mémorial de chronologie, de biographie, d'economie politique, etc.*

over in silence our faithful servants—Mrs. Jankowska and Mrs. Anna Sosnowska who took care of all of us. We spoke Polish with them. In fact, we knew that language better than Russian, so that on our arrival in Petersburg we had great difficulty in getting rid of our Polish accent and Polish sentence structure.*

Thanks to our acquaintance and cordial relations with the people of influence who enjoyed the hospitality of my father, he succeeded in placing almost all his children in the best schools and at the expense of the government. Thus as early as 1827 my brother Alexander was already enrolled in the Noblemen's Boarding School connected with the Imperial Lyceum of Tsarskoe Selo. The following year my sisters Emilia and Valeria were placed by my mother in the Institute named after Catherine the Great. Through the kindness of the Grand Duke and Heir Presumptive Konstantin Pavlovich, I was also placed on the list of candidates for the Lyceum Boarding School. My parents planned to send me there in a few years, when, at the end of December, 1828, in a most unexpected manner, my father received a communication from the director of the Lyceum, Major-General Gol'tgner, requesting that I be brought to the Boarding School because by imperial order I was to be enrolled at the expense of the government as of January 1, 1829. This news embarrassed my parents quite a bit. I was then only eight years and a few months old. I was not particularly robust, and, as everybody told me, spoiled by my mother who had a particularly tender feeling for me. Moreover, that winter was severe, and so long a journey could be too hard for me. Fearing, however, to miss the appointed date and thus perhaps also to lose the right of admission to the Boarding School at government expense, my father decided to send me to Petersburg at once. A favorable opportunity to do so presented itself. One of the customs officials, Kassian

---

* I also recall with particular pleasure our old servants: Anton, who sang Polish songs to me which are preserved in my memory to this day, and our gardener with whom I picked fruit in our large garden. I remember also our lackey, Alexei, and our general concern and grief when he was recruited into the army. No less memorable to me is our Russian coachman, Matvei, whom I consulted often about my horse. I must mention also my faithful companion, our magnificent Danish dog, Pluto, much beloved by us children.

Kassianovich Iankovskii, had to go to the capital to enter the Department of Foreign Trade. I was entrusted to his care by my parents who paid half of our traveling expenses. I remember distinctly the circumstances connected with my departure from Radzivilov. This was the first important event in my life. However, because of my extreme youth, only the superficial aspect of it interested me. Everybody at home began to be occupied with me alone. My mother's love for me was expressed in the tenderest caresses. With tears in her eyes she made all preparations for my long journey. For me, however, because of my foolishness, these preparations were a source of amusement. I was particularly happy about the uniform coat which I was to wear. I was simply enraptured by its red collar! On the eve of my departure one of the post office employees, Dombrovskii by name, a religious man and sincerely devoted to my father, took me to the Catholic Church where the priest read a parting prayer and put a small gold cross on a green ribbon on my neck, with which my mother blessed me. All this is as clear in my memory as if it had happened a few years ago. Finally the fateful day arrived. The tender farewells of my parents, the crying of those present, and the sobbing of the servants—all this had a strange effect on me. I did not realize what was happening to me and began to understand my situation only on the second or third day of our long journey.

Thanks to the magnificent sleigh road and our courier credentials, our sledge tore along like an arrow. We rode day and night, and I recall that several times our courier troika carried us so fast that the coachman could hardly hold the horses back. I was in a daze, unable to understand what was happening to me. In about four days I was in the home of my uncle, Baron Fedor Fedorovich Rozen. There, after I recovered my senses, I felt a tremendous longing for my parents' home. In spite of all the tenderness and the consolations of my kind aunt, Elisaveta Petrovna, and also the amusements that she tried to show me in Vilno, which, after Radzivilov, seemed to me to be a magnificent city, I cried frequently thinking of my parents. I recall that once my aunt played a Polish song which I had often heard sung by my mother. I cannot express the impression this music made upon me. Its sounds were so deeply engraved in my memory that even now I often recall them with sadness and delight. After our stop in Vilno we set out again on our journey during a terrible frost. We stopped for a day in Pskov at the home of one G. Kavetskii where we were cordially received. He was an old friend of my father's and, I think, a wealthy landowner. His house amazed me with its rich and elegant furnishings.

We arrived in Petersburg the latter part of January, 1829.

II

THE NOBLEMEN'S BOARDING SCHOOL
(1829–1832)

I REMEMBER THE DAY of my arrival in Petersburg very well. We stayed, of course, at the home of the Sul'menev's. They lived then on Vasilii Island, on the Eighth Line, between the Embankment and Bol'shoi Avenue, in the house of the Alexander Nevskii monastery. We arrived around eight o'clock in the evening. My uncle and aunt were not at home. They were visiting a friend of theirs, Fedosia Iakovlena Titov.* I was so bundled up that I could not move, and I was carried half asleep from the carriage into the house. There sitting around the tea table were Aunt Roza Petrovna, Marfa and Alena Nikolaevna Brown (two English women, spinsters, who were staying with the Sul'menevs), Nadin'ka, the elder daughter of the Sul'menevs, and Natasha, the youngest, who was about six years old. The other sisters, Katin'ka and Anneta, were then at the Patriotic Institute, and their brothers Nicholas and Peter were in the Naval Cadet School.

They all rushed up to me, put me on the table and began to unbundle me. There was no end to their caresses and embraces. I felt that I was with my relatives and at once felt better. On their return home Ivan Savvich and Natalia Petrovna also showed tenderness toward me. My kind companion, Iankovskii, and I were assigned a room each, and we were glad to go to bed after quite a trying journey.

On the following day my kind aunt, Natalia Petrovna, took me in a sleigh to my uncle Alexander Karlovich, who also lived on Vasilii Island. I was struck by his resemblance to my father: the same features, the same manners, and even the same voice, except that my uncle was somewhat taller. I was, of course, received with open arms in his family.

* At that time she was a widow. Later she married Chief of Staff Major General Vashutin. She was an intelligent and beautiful woman.

I was enchanted by my first ride through the city of Petersburg on a beautiful winter day. And I marveled at the beauty of our capital and the great activity on the streets. I also recall that we encountered a detachment of the Paul Regiment in full uniform. I liked the red plumes on the soldiers' helmets very much.

On the first Sunday my aunt took me to the Catherine Institute to visit my sisters Emilia and Valeria who were being educated there. I cannot describe my joy at this meeting. The trip over the Isakiev bridge,* the Isakiev Square and through the Palace Square and along the best part of Nevskii Prospect thrilled me beyond words, and I constantly questioned my aunt who satisfied my childish curiosity with the same patience and kindness toward me which never changed on her part. I can honestly say that I always sincerely appreciated this and my feeling of love and devotion to this worthy, intelligent, and highly educated woman never changed either.

I grew no less attached to my uncle Ivan Savvich. He was a kind-hearted father of his family with all the remarkable qualities characteristic of a Russian man. Entertaining was a passion with him. All the relatives, even the most distant ones who came to the capital from the provinces, used to stop at his home for weeks and were given a most cordial welcome. It is remarkable how he was able to entertain so lavishly with his small means. To be sure, as the Russian proverb has it: *izba ego byla krasna ne uglami a pirogami* [In his house it was not so much the plate as what was served on it]. Nevertheless, it must have been hard for him to make ends meet with his large family and living in this manner. Personal interest did not exist for him. He denied himself even the slightest pleasures, but he was constantly concerned about others. Ivan Savvich was the model of a real Christian, and his piety was completely devoid of bigotry. He was simple in his relations with everyone, and he disliked all formalities. Following the old custom he always addressed those younger in years than himself by the friendly *thou*. Although he was not highly educated and knew no foreign languages, he was, nevertheless, not without knowledge in the

* The Nicholas Bridge was not built until fifteen years after my arrival in Petersburg.

sciences, particularly in his own field of specialization as a seaman. He was intelligent and endowed with remarkable common sense.

When I arrived in Petersburg Ivan Savvich was more than sixty years old. He was tall, stout, and quite awkward. His features were coarse, of the Tartar type, but his face became kind and tender in conversation. He was extremely quick-tempered, but his momentary wrath did not frighten anyone because he immediately became as gentle as a lamb. This happened frequently when he played cards. The slightest mistake used to enrage him. "You fear no God," "You never studied logic," he would shout at his unfortunate partner. In his anger he would sometimes reach for a candlestick, but suddenly he would calm down and his thunderous laughter would be heard throughout the house. And it should be remarked that the game almost always was not for stakes, except perhaps for kopeks. He loved to pass his evening this way and played for the most part with his relatives or with the children.

Ivan Savvich was a full admiral when he completed his service. However, the title of His Excellency in no way changed his modest and simple relations with his juniors. After leaving active service in the fleet in the rank of captain first class, he was for a time inspector of the Naval Corps, then after his promotion to rear admiral, he was director of the Auditoriat Department for a long time (he held that title when I arrived in Petersburg), and, finally, member of the Auditoriat Council.

My most estimable uncle was a nobleman from the province of Riazan. He owned a small estate there, but with his characteristic kindness and selflessness, he placed it at the complete disposal of his brother, Alexei Savvich (who resembled him in no way morally) and his sister. The only good that he derived from his estate was the fact that his servants were sent to him from there. His entire staff of servants consisted of serfs. This was entirely in keeping with the custom then prevailing. The small two-story house he occupied was not outstanding either for its beauty or its elegance. But it did not represent a rarity in Petersburg because of its many conveniences for family living. What things did it not hold!

In the Sul'menevs' house I met all of our relatives who then lived in Petersburg. But I did not remain there long this time. I had to be presented quickly at the Lyceum Boarding School, and I was taken to Tsarskoe Selo.

I went with great pleasure to Tsarskoe Selo, knowing that I would meet my brother Alexander whom I had not seen for several years. However, on my arrival this feeling was completely overshadowed by awe after my presentation to the authorities and by the formal appearance of the class. I was placed, of course, in the lowest class where I was the smallest and practically the youngest pupil. At the end of the day I experienced all of the fears of a beginner. I was overwhelmed by the questions and the unceremonious pranks of my countless classmates. Students from all the classes filled the recreation hall and raised a terrible racket. I was completely bewildered. Observing my pitiful state, the tutor on duty (Kokorev), took me under his wing. Soon a bell was heard calling us to supper. We were led into the dining room. It was an enormous room that easily accommodated all the students of the boarding school. There were over two hundred of them! They sat in groups according to classes at large tables. Never had I been at such a gathering, and the impression it made upon me was most unpleasant. I was frightened.

After general prayers we were marched to bed. Each student had a separate small room on the order of a cabin, which contained a bed and a dresser. In cases where there were two brothers in the Boarding School, they were placed together in a somewhat larger room. This was the case with us. Left alone with Alexander, he began to encourage me and to calm me. In a few days I became completely accustomed to my new environment and even enjoyed myself because everybody liked me and the administration even spoiled me because of my young age and good behavior.

The Lyceum Boarding School was housed in a large and beautiful building in a suburb known as Sofia, very near the park of Tsarskoe Selo, from which it was separated only by a wide street.* Adjoining the house, in the back, was quite a large birch grove where we used to play in our free hours. The barracks of the Household Hussar Regiment were practically next to us, and we often

* At present, the Rifle Battalion is housed there.

heard the lovely music of this regiment. The handsome hussar uniforms and their magnificent horses were a common sight to us. All this added great excitement and gaiety to our surroundings. As far as I can remember, living conditions were superior in our school to those in other secondary institutions. Our building was excellent, we were well fed, and the administration treated us well. It seems to me that the teaching was also of a high quality because our instructors were for the most part members of the Lyceum faculty. Our Boarding School was part of the Lyceum.

At the same time, the director of the Lyceum was also head of the Boarding School, which, however, had its own director. In my time this office was held by Colonel Vaksmut.

Our school served as a preparatory school for the Lyceum. Such was its prime purpose. At that time the Lyceum of Tsarskoe Selo, which was justly regarded as the best institution in Russia, consisted only of two sections, each comprising twenty-five boys. There was no other division into classes. Each group remained in the Lyceum for three years. Consequently the entire education in the Lyceum lasted six years. Every three years after the final graduation, students from the junior course passed into the senior course and were replaced by twenty-five new students, selected by examination from the best Boarding School pupils, aged twelve to fourteen and a half years.

Our Boarding School had seven classes, and the total enrollment was about two hundred boys; and, since admission was possible only every three years and at a certain age, very few boys had access to the Lyceum. All the others continued their studies through the seventh class. The best of these graduated with the rank of collegiate secretary (tenth rank) in the civil service. And those who chose a military career were given the rank of cornet in the Young Guard.

All this was very well planned. The noble spirit of the Lyceum also prevailed in the Boarding School. Both of these institutions comprised one unit. Education in peaceful Tsarskoe Selo, far away from the tumult and temptation of the capital, brought excellent results as is well known. Not only the graduates of the Lyceum, but also the students of the Boarding School made a name for themselves in their chosen fields as useful citizens and faithful sons of

Russia. One naturally regrets the closing of this institution founded by Alexander the Blessed!

The august founder of the Lyceum had the intention of creating a perfect model institution where he intended to send his younger brothers, the Grand Dukes Nicholas and Mikhail Pavlovich, to be educated. He, consequently, wanted to give them worthy companions. The dowager empress was opposed to his plan. Nevertheless, the Lyceum was founded in 1811 on the above-mentioned basis and was housed in one of the wings of the palace of Tsarskoe Selo, and the Boarding Schol connected with it was set up in Sofia. Both of these institutions were placed under the jurisdiction of the tsarevich, Grand Duke Konstantin Pavlovich. The tsar himself also took a personal interest in it and frequently visited the Lyceum, which was separated from the chambers of His Majesty only by a corridor which served as a library. The tsar knew each student of the Lyceum by name.

During the last years of his reign Tsar Alexander Pavlovich, as is known, abandoned somewhat his liberal ideas and fell under the influence of Arakcheev. This change had its repercussions at the Lyceum also. In 1823 the unforgettable director, Egor Antonovich Engel'gardt [Engelhardt], under whom the Lyceum had so prospered, was replaced, and his post was taken by General Gol'tgner, a man acquainted only with military service who was totally unprepared for his new calling. This was the first blow for the Lyceum, which bitterly mourned the profoundly respected and beloved Engel'gardt. However, General Gol'tgner was not a bad man. He was moderate in introducing the strictness and discipline which were expected of him when he was entrusted with the administration of the institution.

Affairs fared even worse after Emperor Nicholas ascended the throne. In 1829 it was decided to discontinue the Boarding School connected with the Lyceum.

The cause for so severe and unfortunate a measure was some boisterous pranks that took place in the upper grades of the Boarding School. But could not a measure of punishment and correction have been found other than the final closing of so useful an institution? We were informed of the lamentable news that the Boarding School was closing shortly after Easter. We were, in addition, told by the administration that the students would be distributed throughout various other institutions. I was assigned to the Noblemen's Boarding School connected with St. Petersburg University. My brother, however, entered the Lyceum since that was the year when a new group of the best students of a given age was being recruited from the Boarding School. My brother was among the twenty-five chosen.

Thus I remained only a short time at the Lyceum Boarding School, which made such a pleasant impression upon me. The magnificent building of our Boarding School was made available for the Cadet Corps for small boys. Was this not the real reason for our banishment from Tsarskoe Selo?

I returned to Petersburg during my summer vacation, which I spent with the Sul'menevs. They lived then in the dacha of Skorodumov on Aptekarskii Island, right on the shore of the Nevka, opposite Kamennyi Island.

I liked the dacha life on the islands very much. It was much more lively then than it is now. The court passed part of the summer on Elagin Island. The grand duke Mikhail Pavlovich at that time also lived in his magnificent palace on Kamennyi Island. From this spot on there were a series of beautiful and sumptuous dachas which then belonged to the best aristocratic families— Dolgorukii, Bobrinskii, Gagarin, and others. The excellent band of the Mounted Guard Regiment, stationed in Novaia Derevnia, played every evening in front of the Elagin Palace. Public outings, magnificent celebrations with illuminations and fireworks regularly presented from the dachas of the aristocracy added unusual excitement to the islands. Tsar Nicholas Pavlovich was still very young then and at the peak of his handsomeness. Empress Alexandra Fedorovna, a striking and exceedingly graceful woman, was a worthy companion for him. The imperial couple loved luxury and amusement and were surrounded by a highly impressive court. Strolling with my uncle Ivan Savvich, we would frequently encounter enormous cavalcades, headed by the handsome tsar with the empress. This sight, so unusual for me, used to fill me with enthusiasm. Often we would direct our steps toward the park of Count Laval on Aptekarskii Island, right next to our dacha. This magnificent park, facing the Little Neva and the Karpovka, is memorable to me because it was the place of our childhood games. I recall that during one of our walks we heard a tremendous cannonade from the Fortress of Peter and Paul. This was on the occasion of the conclusion of the Turkish war and the signing of the peace of Adrianople.

On Sundays and holidays we used to go to the modest church on Kamennyi Island. I grew to love the services of the Russian Orthodox Church very much. Watching my kind and estimable uncle pray—and he always prayed aloud and with great feeling— I too was filled with a Christian feeling which, I must say, became deeply and forever rooted in my heart. From then on I always went to the Russian Orthodox Church to attend mass on holidays, and I did so the more readily because with my limited knowledge of the German language I understood very little of the long sermons in the Lutheran Church. Moreover, because of my sensitive character, the whole setting and the singing in the Russian church predisposed me to prayer much more so than did the service of the Protestants, which was dry and devoid of ritual.

Speaking of our life at the dacha, I cannot help but recall that Ivan Savvich, as an old seaman, completely disregarded the wearing of the military uniform, which was observed with such unusual strictness in the reign of Nicholas Pavlovich. In spite of the presence of the court in our immediate neighborhood, my dear uncle frequently took walks in a long civilian coat and a straw hat. I recall how worried my dear aunt was lest he encounter the tsar in this garb. But, thank God, he never met with any unpleasantness in this respect, despite the fact that he did meet some of the superior officers who knew him personally. They all respected the old man, and knowing his eccentricities they never reprimanded him for not wearing his uniform.*

In the month of August at the end of my vacation I was taken to the Noblemen's Boarding School of the university which was situated on Ivanov Street near the Semenov Regiment. This institution, as its name implies, served as a preparatory school for the university, which was situated at that time on the corner of

---

* During my stay at the dacha this year we received from Radzivilov the news of the birth of my sister Anneta.

the same street, next to us. However, not many of our students entered the university. I do not know whether it was because the number of university students was greatly limited or because our students were satisfied with the education they received in the Noblemen's Boarding School, for the outstanding among them graduated with the rank of collegiate secretary, tenth rank.

Our Boarding School was under the jurisdiction of the Ministry of Public Education and under the direct administration of the chief of the St. Petersburg Educational District. This post was held then by Konstantin Matveevich Borozdin, who visited us frequently and was loved and respected by all. The inspector of our Boarding School, Klemens, enjoyed the same affection. We had no director at the time.

The University Boarding School made an unfavorable impression upon me. Everything about it seemed to me to be inferior to the Lyceum Boarding School. The building, the food, the uniform, the treatment of the students—everything compared unfavorably with that which prevailed in the institution of Tsarskoe Selo. However, I soon grew accustomed to my new life and was happy with my lot. Nevertheless, I always waited impatiently for Saturday to come. On that day after our dancing lesson we were dismissed to go home, returning Sunday evening. I was always called for by Evstafii, a noncommissioned officer, a permanent orderly who had been with Ivan Savvich for more than twenty years as a sort of valet. We usually returned home with Evstafii in a cab, and Evstafii never failed to choose the cheapest, known as *van'ka*, pulled by a miserable jade. In this conveyance we dragged endlessly all the way to the Eighth Line on Vasilii Island. We invariably went the full length of Gorokhov Street. Because many sleighs used this street in winter, the road was very bad, and it was hard for the poor jade to pull us. Our Van'ka would then get off his seat and walk almost the entire way, urging his little horse on. Who among the Petersburg inhabitants is not familiar with such a scene, one of the most unique aspects of our capital? It was even less comfortable riding on wheels. The *drozhki* which were in use in those days and which resembled a double-bass in shape, forced the passengers to remain in a painful position to keep from falling out.

Saturday was memorable not only because of the dancing lessons given by our dear and charming Frenchman, Dutak [Dutaque], but also because immediately following this lesson came the *execution*, or, in other words, the birch flogging of those who had misbehaved during the week. This form of punishment was used widely then, and not only for young pupils, but also for students in the upper grades. Those guilty of a minor offense were forced to be present at the execution as a warning. Once that happened to me, and I had to be among the latter because of some minor prank. This made such a strong impression upon me that on returning to the Sul'menevs, I was unable to say a word during the entire evening.

However, this was the only punishment to which I was subjected during my entire stay in that institution. Without boasting, I can say that I was always distinguished by my good behavior in spite of my cheerful disposition. The bad and vulgar pranks of some of my classmates only aroused loathing in me. My classwork was fairly good also and the report card I had to show at home was always marked "good."

I used to spend my weekends pleasantly in the family circle of the kind Sul'menevs. On those days all of their relatives gathered in their home, and there were always a great many children. Ivan Savvich used to take in for the holidays not only his relatives from various institutions, but also the children of his friends who had no family in Petersburg. There were often more than twenty people at the table. The old man was very hospitable, and he was heartily happy to see this large gathering. Moreover, Ivan Savvich had no difficulty in providing sleeping accommodations for us young people. Hay was spread on the floor of his study, covered with sheets, and thus each one of us had a bed where we slept soundly after playing and running around all day. In the morning my honorable uncle used to come to the same room to dress. Here too in front of the icons he would loudly say his warm prayers to which we always listened reverently.

On holidays his sons Nicholas and Peter * had to report to the Naval Corps of Cadets for the ceremony of the changing of the

---

* I was especially friendly with Peter. The older brother was somewhat willful and hard to get along with.

guard. I almost always accompanied him to church, then I used to visit my sisters at the Catherine Institute with my aunt, and sometimes we used to go also to the Patriotic Institute where the Sul'menevs' daughters were being educated.

While reminiscing about the time spent in this family, which was so dear to me, I must mention the two elderly English women, Marfa Nikolaevna and Alena Nikolaevna Brown, who were always sincerely kind to me and showed almost a parental interest in me. Their father, a colleague of Ivan Savvich, died and left his children orphans in Russia. My kind uncle gave the two girls a home in his house, where they remained for the rest of their lives. Their older brother, Nicholas Nikolaevich Brown, attained the rank of colonel and was commander of the Petersburg Gendarme Division. Two younger brothers, Egor Nikolaevich and Ivan Nikolaevich, were naval officers. All of them visited the Sul'menevs often and were treated like relatives.

My mother's younger sisters, Emilia and Roza Petrovna Litke, were at that time in the full bloom of their beauty, and both of them were of an unusually gay disposition. We particularly loved Roza Petrovna for her remarkably kind heart. She spent all of her life looking after and caring for those near to her, never thinking about her own personal interests. My mother's older brother, Evgenii Petrovich, who held some sort of post at the Academy of Science (I think he was in charge of the printing office) used to be a daily visitor at the Sul'menevs also. Although remarkably gifted and kind-hearted, he was, however, extremely irritable and because of this was, of course, unable to make a place for himself either in his work or in society. Although sincerely fond of the Sul'menevs, he was constantly quarreling with them. I recall that on one occasion as a result of some trivial argument at the dinner table, he became so excited that he left the table and did not come to the house for several days. Then invariably a most touching reconciliation would follow. He was a connoisseur of music and had such a remarkable musical ear that after hearing an opera only once he could improvise on the piano some of the principal arias. He was equally gifted in foreign languages, of which he had an excellent command. In his youth he had taken care of the foreign correspondence of Senator Krasno-Miloshevich when he ruled

Moldavia during the Turkish war of 1828. As I have already said above, my uncle Evgenii Petrovich died around 1832.

Another brother of my mother's who was in Petersburg was Nicholas Petrovich. In disposition he greatly resembled Evgenii Petrovich, but he could by no means equal him in education. After serving for several years in some army infantry regiment, he entered the civil service in the Department of the Holy Synod. At that time he was still a bachelor, and we used to take walks together. During Mardi Gras and Holy Week I visited the show booths with him which amused me greatly.

At the end of 1829 my uncle Fedor Petrovich Litke returned from his second voyage around the world with his brig "Seniavin," which he commanded. His arrival was the occasion of general rejoicing in our family, and all of us marveled at his success in the service. The tsar Nicholas Pavlovich received him very kindly and promoted him from lieutenant-captain to captain first grade, which meant that he skipped one rank. Fedor Petrovich was then presented to the empress to whom he had to recount his interesting voyage. His enlightened mind, remarkable gifts, and fine manners earned him particular attention on the part of the tsar, and he soon afterward was made aide-de-camp to His Majesty—a rare honor in those days, particularly with regard to seamen. Fedor Petrovich earned even greater attention in the world of science than at court. The famous Humboldt [1] and Cuvier [2] entered into correspondence with him, praising his scentific achievements and discoveries. The Greenwich Observatory made him a member for his explorations in the Aleutian archipelago, where he discovered an island which he named "Seniavin." However, Fedor Petrovich had already earned great fame as a navigator following four of his voyages to Novaia Zemlia, whose shores he surveyed very carefully. These voyages entailed great danger.

Bound by a great friendship with his sister Natalia Petrovna who actually worshipped him, Fedor Petrovich spent the part of his time which was free from studies in the home of the Sul'menevs, where he was the center of all activities. In spite of the serious bent of his mind and his interest in scientific work, he had a very cheerful disposition. He liked to joke, and his loud laughter was often heard throughout the house. He rented an apartment for himself,

two steps from the house of the Sul'menevs, and I visited him frequently because he made me come with my books to prepare my
lessons during my vacation. My uncle liked me very much and his
kind attitude toward me never changed and remains the same to
this day.

I grew so attached to all the relatives on my mother's side that
I visited unwillingly and quite seldom my father's only brother
who lived in Petersburg, Alexander Karlovich Giers. However, I
did love him very much. The reason for my reluctance to visit him
was that I did not have too much in common with his friends who
used to gather in his home. Moreover, the conversation there was
for the most part in German which always embarrassed me. Fedor
Petrovich Maiet was among my relatives on my father's side whom
I used to meet there. He was a seaman and was serving on the ship
of my uncle, F. P. Litke. However, he did not complete the voyage
around the world with him. For some reason he incurred Litke's
displeasure, who transferred him to another ship they met, a Russian battleship. I also used to meet there Konstantin Petrovich
Maiet, a most charming and remarkably handsome young man,
highly gifted and an excellent musician.

In keeping with the German tradition Alexander Karlovich always had a tree on Christmas Eve, and I was invariably invited
there. His children, whom I mentioned at the beginning of my
notes, were at that time still very young and I was not as close to
them as I was to the children of the Sul'menevs. I have already
spoken of the attractive character of Alexander Karlovich. His wife
Elisaveta Karlovna was also the kindest of women and so was her
sister Louiza Karlovna Tiffenbakh, an old spinster who lived in
their home and who devoted her entire life as well as her fortune
to her sister's family. However, this fortune too did not remain
intact in the hands of my dear extravagant uncle. I must also mention my aunt Emilia Karlovna Tiffenbakh, an old busybody, always worried and anxious about the future of her niece's large
family, who often really were in a difficult situation because of the
irresponsibility of the kind and impractical Alexander Karlovich.
Nature was unkind to her. Of short stature, with a large hump and
an unusually long face, this restless and quick-tempered old woman
often made us laugh. But these faults of hers were more than com-

pensated by her spiritual qualities and her remarkable mind. She sometimes visited the Sul'menevs where she was appreciated.

I received frequent news from my parents, either directly or through my aunt Natalia Petrovna, who was in constant correspondence with my mother. In one of the letters, which I have preserved to this day, my father notified me that my little horse which I used to ride so often in Radzivilov had died. This grieved me very much. My mother wrote me quite often. Never ceasing to miss me, she expressed the desire to have a portrait of me and wrote about it to Natalia Petrovna. In those days people did not have the slightest knowledge of photography or daguerreotype. That meant that an artist had to be found. This task was entrusted to Alexander Fillipovich Postel's,* who, in the capacity of a naturalist, had been a member of the expedition of Fedor Petrovich which went around the world. The artist whom he brought executed the portrait satisfactorily as far as I have heard. The portrait he painted, representing me in the uniform of the University Boarding School was passed on after the death of my parents to my sister Iulia, who presented it to my wife during our meeting in Kishinev in 1854. Thus it is now in our home.

The events of the years 1830 and 1831 are very clear in my memory: the insurrection that flared up in Warsaw and the Polish campaign which followed were a subject of conversation even in our Boarding School. We read the reports on military actions with interest and were fired with a militant spirit. The departure of the Guards for the front presented an interesting spectacle to us. However, our attention was soon directed toward a different event. Rumors spread concerning the appearance of the epidemic, unknown until then, which was carried to Russia from Asia—*cholera*, or as it was called then, *cholera morbus*. Stories were told with horror about the devastation caused everywhere by this epidemic and about its approach to Petersburg. Precautionary measures against the infection were taken in our Boarding School. The rooms were constantly being fumigated with chlorine. All students had to wear a bag with garlic around their necks. The diet was changed.

* A. F. Postel's, an intelligent and educated man, was later director of the Petersburg Second Gymnasium.

Fruit and our favorite beverage, *kvas,* were forbidden. All this fuss amused us children, but our instructors were sad and dejected. The effect of this upon us was that the administration treated us in a more lenient way, or perhaps it was just indifference. They had other affairs on their minds. Finally, the disease appeared in Petersburg. Following this announcement, all students were immediately dismissed to go home. Since I had for some time had no opportunity to inform the Sul'menevs about this and since they had already left for the dacha, I had to remain about two weeks longer in the Boarding School with a few other students who were in the same situation as I, or who did not have any relatives in Petersburg.

This brief period remains very vividly in my memory. The classes were discontinued. We roamed around the yard the whole day long or looked out of the windows to see what was going on in the streets. The city was in the grip of a great panic. The public, unable to understand the terribly high mortality and attributing it to poisoning, began to become restless. The Poles were suspected of allegedly bribing doctors to poison the water. There were many victims of this insanity. Once from our window we saw a butcher leave his stand and attack a poor passer-by in a velvet overcoat, whom he dragged by the collar shouting in rage that he saw him throw some powder into the water in order to poison it. Everybody knows about the riot on Haymarket Square during these unfortunate days which was suppressed by the remarkable courage and presence of mind of the tsar.[3] It was then too that the enraged mob moved on to the Maria Hospital, destroyed everything there, smashed the windows and tore to pieces the unfortunate doctor, Evropeus. Screams were also sometimes heard in our ordinarily quiet Ivanov Street, and we were in a constant state of alarm. The disease penetrated our Boarding School also. Our French tutor Badion and several maintenance men died. In the midst of these events the appearance of Evstafii, who finally came to take me to the Sul'menevs, cheered me greatly. When riding with him to Aptekarskii Island, we passed the recently destroyed Maria Hospital, which was a very sad sight. The streets seemed quiet. Thanks to the wise measures taken by the government, the mob was calmed everywhere; moreover, the epidemic began to subside. It continued to the end of autumn, but in a much milder form. Since this first and terrifying appearance of the disease, it frequently re-appears in our capital where the people are already so used to it that they practically pay no attention to it.

My summer vacation was spent very quietly in the Skorodumov dacha. At that time the sons of the Sul'menevs, together with the rest of the cadets, were taking part in the Mirsk maneuvers, an event which henceforth was repeated annually for them at vacation time. Fedor Petrovich Litke was delegated to accompany the Grand Duchesses Maria, Olga, and Alexandra Nikolaevna to Reval and to remain with them during their cure. Our other relatives seldom came to see us in the dacha. This solitude brought me even closer to my aunt Natalia Petrovna, who treated me like her son. The instructive talks with this intelligent woman interested me very much. Moreover, my playmate at that time was her youngest daughter, Natasha. I recall that once on entering the living room I found my aunt in tears. On my inquiring as to the cause of her grief, she pointed to the organ grinder who at the moment was playing outside the window, and said; "Look, my dear, at this un-fortunate man, who for the sake of his daily bread must trudge all day long in this heat with this heavy instrument. He plays jolly tunes, but who knows what sorrow he has in his heart and whether tomorrow he will not be a victim of cholera." I was always very sensitive and fully shared the sad mood of my aunt.

During the summer of 1831, if I am not mistaken, my uncle Baron Fedor Fedorovich Rozen came from Vilno with his wife Elisaveta Petrovna and stopped as usual at the Sul'menevs. The purpose of their visit was, as I remember, to place their elder daughter, Matilda, in the Catherine Institute, where my two sis-ters Emilia and Valeria were studying. I think that I have already mentioned that my mother and Elisaveta Petrovna had received their education at the Institute, but their elder sister Natalia Pe-trovna Sul'menev was educated at home.

In the summer of the same year the funeral service of the tsare-vich, Grand Duke Konstantin Pavlovich, was held in the Peter and Paul Fortress, which was not far from our dacha. He died of cholera almost at the very beginning of the Polish campaign. As is known, everybody in Russia was dissatisfied with his actions both as a commander and as viceroy of the Polish kingdom. As a result the news of his death was generally received with some joy. In spite of a pouring rain, we went, I do not recall with whom, to witness the funeral procession. The tsar followed the coffin on horseback, and he was completely soaked. Water dripped from his Andrei ribbon on to his white buckskin breeches and turned them a bluish color. And yet I could not help but admire the majestic sight of the tsar.

Around this time important changes took place in our Boarding School. With the appointment of Uvarov (subsequently made Count) to the post of minister of public education in place of the infantry general Baron Khristofor Khristoforovich Lieven, the secondary schools in Petersburg were reorganized into gymnasia. Our University Boarding School was the first; the so-called High School was the Second Gymnasium. Then the Gymnasium proper was designated as the Third, and the Larinskii School as the Fourth Gymnasium. Our institution, from which hitherto one could graduate, just as one did from the Lyceum Boarding School, with the rank of collegiate secretary, tenth rank, lost this privilege and had to submit to the same regulations which applied to all gymnasia. We naturally were greatly displeased with these changes and accepted grudgingly the designation of gymnasium students, which was then pronounced with a certain contempt.

At that time I was already in the second class, and my studies were proceeding very well. In fact, I was rated as one of the best students both scholastically and from the point of view of good conduct, because I disliked taking part in the boisterous and naughty pranks of my comrades, some of whom, I must confess, were a pretty bad group of boys. I was closer to the more modest and well-bred among my comrades. Among them Konstantin Veselovskii, an unusually gifted boy, stood out particularly. My bosom friend was Obolenskii, who soon left the gymnasium to serve in the navy. Among my friends of that period I should like to name Burtsov, the son of a landowner of Riazan, a neighbor of my uncle Ivan Savvich (because of this he frequently spent his holidays in the Sul'menevs' home); also Brok, who chose the same career as Obolenskii. He now holds the rank of rear admiral and is a member of the retinue of His Majesty. He also fulfills the duties of *Hofmeister* for Grand Duke Vladimir Alexandrovich.

Finally, I should like to mention Perets, an enthusiastic youth who studied poorly and instead spent whole days reading the patriotic novels of Zagoskin, reciting poetry, and drawing very clever sketches of his heroes. He sat next to me in the classroom and I owe to him the fact that I read many books. I recall my first book *Bednaia Liza* (*Poor Liza*) by Karamzin, over which I shed many tears. I then read avidly *Iurii Miloslavskii* and *Romashev*. The reading of these novels made the profoundest impression upon me and developed my sense of patriotism to the highest degree, a feeling, which, thank God, I still preserve to this day despite the fact that I have been living in foreign countries since the age of twenty-two.

Among the students in the upper grades, I recall Vladimir Il'ich Vestman, now undersecretary of foreign affairs, Iazykov, Panaev,[4] who became famous in Russian literature, and the two brothers Balabin. The elder of them became ambassador in Vienna, and the younger, Evgenii, a charming and quiet young man, for some unknown reason fell under the influence of the Jesuits, became converted to Catholicism and took monastic vows in some monastery in France.

With the changes that followed in our school, Benard, a colonel in the Transportation Corps, was appointed director of the Gymnasium. I personally cannot complain about him because he favored me very much. Our inspector was one Bashinskii, whom I never recall seeing relaxed. He exercised his duties with unusual zeal and passion. He was constantly reprimanding somebody, and one could hear his voice everywhere.

Among the instructors I recall particularly Bardovskii, a teacher of Russian; Popov, a teacher of history, beloved by everybody, a very gifted man, who shot himself in a fit of melancholy, a fact which shocked all of us and also grieved us; and our teacher of religion, Reverend Vasilii Bazhanov, endowed with all the qualities of mind and heart fitting to his calling. His appearance was also very attractive. He conducted the services with great dignity and feeling, and it was a genuine pleasure for me to attend the university church where he always conducted the services. In addition to all of his gifts, he also had an unusual oratorical ability. His sermons were delivered with great enthusiasm. All these qualities attracted the attention of Tsar Nicholas Pavlovich, who made him the chaplain of the tsar's family. Father Bazhanov has held this rank for the last forty years and he enjoys general respect.

As I said above, the epidemic of cholera continued into late fall. After the Sul'menevs returned from the dacha, I continued to visit them in the city during my holidays. Once Ekaterina Fedorovna Korostovtsev, my mother's cousin, was among our guests at the tea table. The conversation was about the precautions necessary against the epidemic. Ekaterina Fedorovna argued against the need of such precautions and began to eat raw fruit right there and then at the table, ridiculing those who did not care to follow her example. The following morning one of her man servants came running to us with the news that she had contracted cholera. She died that same evening. It is easy to imagine how this sad news affected all of us. As I have mentioned above, I used to meet the Korostovtsev children often in the home of Sul'menev, and, although they were all much older than I, we were great friends.

Among the victims of cholera in 1831 I must also mention our famous navigator, Vice-Admiral Vasilii Mikhailovich Golovnin, who was related to Ivan Savvich and visited his home frequently. Subsequently, I became very friendly with the family of Golovnin under circumstances about which I shall speak later.

If I am not mistaken, this happened in the same year that we received the news that my older sister, Sofia, had married Collegiate Councilor Paul Sitnikov, who, in connection with his official duties, was stationed in Dubno, near Radzivilov. Sitnikov was a widower with several children from his first marriage. Two of his sons were brought to Petersburg to be enrolled in the Naval Corps. And as was to be expected, they were received by the kind Sul'menevs as if they were their own children. They always spent their holidays with us. The younger of them died shortly after his arrival, and the elder, Paul, later turned out to be a very good naval officer in command of some battleship. I have now completely lost touch with him. Sitnikov also had a daughter Vil'gel'mina Pavlovna, who married Major Kalinkovskii. After his death she moved with her two children to his estate near Petersburg which she inherited. Vil'gel'mina Pavlovna's daughter married Doctor Kapella whom we shall meet later. My sister Sofia died in

Dubno a few years after the death of our father, leaving behind a daughter, Elena, who died in childhood.

I recall also that at approximately this time (I am speaking of 1831) an unknown young man paid the Sul'menevs a visit one day. He was of a very pleasant appearance and was introduced to the relatives with some ceremony. It was explained to me that he was engaged to my aunt Emilia Petrovna Litke. His name was Vasilii Matveevich Maresev. He was then secretary in the Senate, but just before his marriage he was transferred to the Ministry of War and became the head of the division of the Engineering Department of the War Ministry. I was not present at his wedding because it took place on a week day when I was in school. The Maresevs moved into a beautiful government apartment in the Engineering Quarters (Mikhail Castle) where they lived almost thirty years. The view from the windows of their apartment on to the Summer Garden and the Tsaritsyn Meadow was lovely. We visited the Maresevs, who were very hospitable, and every time there was a parade on the Tsaritsyn Meadow, we went to their apartment to admire the magnificent military spectacle.* God blessed the Maresevs with an enormous family.

After the marriage of Emilia Petrovna, her younger sister, Roza Petrovna, the only remaining member of the Litke family, lived with her mother, Ekaterina Andreevna, now Zavalishin by her second marriage. Her life there was not an enviable one. In spite of her kind and gentle character, the Zavalishins did not like her. Their entire attention and concern were centered in their only daughter, Elisaveta Afanasievna, a very pretty, but spoiled, little girl. She later married an engineering corps officer, Pasypkin, who, after an extended service in the Caucasus, returned to Petersburg with the rank of general. Suffering many injustices and much unpleasantness at home, the gentle Roza Petrovna began to lead a completely nomadic life. For days and even weeks she would live

---

* The Maresev's apartment, however, suffered from one great drawback. In order to reach it one had to go up a stairway of more than a hundred steps and pass through the dark corridors of the castle. I always passed with trepidation one grim place, a tremendous door with an iron bar. It led into the room where Emperor Paul was assassinated, and it was therefore always locked. They say that this room has been opened during the present reign.

now at the Sul'menevs, now at the Maresevs. She also frequently visited her aunt on her mother's side, Roza Andreevna Bolgarskii, the wife of the then prominent senator who was regarded as a very intelligent and business-like man.

The Bolgarskiis lived quite lavishly in their own home, right on the embankment of the Neva, on Vasilii Island, near the Ninth Line. And the Zavalishins lived near Kharalampov Bridge. I often called on both families with the Sul'menevs and recall being quite bored there.

Early in 1831 Uncle Fedor Petrovich Litke left us for a considerable length of time. He was commissioned by the tsar to go on a very important assignment to Danzig in connection with the following matter. With the appointment of Paskevich to the post of commander in chief to replace the deceased Field Marshall Dibich (Diebitsch), who replaced Tsarevich Konstantin Pavlovich, the Polish war began to take a turn for the better for us, but the army felt the lack of supplies. Together with the well-known diplomat and economist Tengoborskii, Fedor Petrovich was ordered to purchase grain in Danzig and to make sure that it was delivered to the army as soon as possible. This commission was executed perfectly, and Fedor Petrovich was honored with the Order of Vladimir third grade. From this time on a close friendship developed between those two intelligent men which lasted to the death of Tengoborskii around 1860.

News from the theater of war grew more favorable daily, and finally the announcement was made that Warsaw had been taken and the Poles subdued. Elevated to the title of Prince of Warsaw and showered with all possible honors by the tsar, Paskevich became the topic of conversation everywhere. I recall that on the occasion of the return of the Guard from the campaign there was a parade on the Tsaritsyn Meadow. Great honors were extended at that time to the field marshal who arrived in Petersburg. All eyes were on him, and when the troops passed in a ceremonial march, the band struck up the Polish national song: "Poland has not yet perished as long as we are alive!" * We all thought it was extremely

* Here are the original Polish words of this song: *Jeszce Polska nie zginęła, póki my żyjemy.*

appropriate and quite in place, but the foreigners did not like it. The English and French newspapers vigorously attacked our tsar for his "ridicule of the conquered enemy" and called this act base and barbarous! In the eyes of the tsar the victory was not over an enemy, but over insurgent subjects.

With the end of the war Nikita Petrovich Pankrat'ev also returned to Petersburg. He was then appointed governor-general of Warsaw. I was attracted by his manly appearance and by his elegant manners. He was, as I have already said, my mother's first cousin, and he treated me like a close relative. Bound by friendship to Natalia Petrovna, who on the death of her father lived with his parents, Nikita Petrovich used to come to the Sul'menevs every day in spite of the very busy life he had to lead during his brief stay in Petersburg. He died about three or four years after his return to Warsaw while still in the prime of life. He was mourned by all his friends and relatives. Nikita Petrovich was a highly educated man, with a kind heart and the most noble, one could almost say, chivalrous character. One of his successors was Prince Mikhail Dmitrievich Gorchakov who held the post of chief of the General Staff of the active army at the same time.

Approximately at this time, or perhaps even earlier, news reached us from Vilno of the death of my uncle Fedor Fedorovich Rozen. After his death my aunt Elisaveta Petrovna moved to Petersburg with her two daughters. And as she was a very well educated and intelligent woman, she soon received the post of *dame de classe* at the Catherine Institute. This institution was headed at that time by Madame Kremnin. From that time on I could visit the Institute much more often and see my sisters not only during the hours reserved for relatives on holidays, but also on any day or at any hour, because I merely dropped in to see Baroness Rozen where invariably I would find Emilia and Valeria during their free hours after class. And I always felt quite at home there. My aunt's eldest daughter, Matilda, was also a student in the Institute. Her younger daughter, Liza, was not yet of school age. In the home of Baroness Rozen I used to meet my old friend, our music teacher, Attse, who on leaving Radzivilov moved with his family to Petersburg. He gave music lessons to my sisters. We used to have much fun with this unique man. Short and homely, Attse could not learn

Russian in spite of his long stay in Russia, and he always confused it with Polish. He was very touchy and short-tempered, however, and would get very angry if we could not understand him or if, God forbid, we would laugh at his way of speaking. One winter morning he came to give my sisters their lesson. No one was in the room, but the maid who was dusting. Attse thought it was cold in the room and going over to the furnace asked: "Tapuli?" [5] The maid opened her eyes wide and indicated that she did not understand. "Tupali?" said Attse raising his voice. The embarrassed maid, not understanding anything, asked him what he wanted. "Tupali, tapuli," shouted Attse pointing at the furnace. Suddenly realizing what he meant, the maid burst out laughing. Whereupon the enraged Attse tried to strike the maid who screamed and raised her brush to protect herself. At this moment Elisaveta Petrovna entered the room and on learning the cause of the incident began to laugh until tears ran down her face. That was too much for Attse who left the room in a rage, and for a long time he did not make his appearance at my aunt's. Such scenes were not infrequent with him. However, he was a very kind man and was greatly attached to our family. As I have already said, he gave his daughter in marriage to the famous Russian pianist Gerke.

In 1831 the Sul'menevs spent their last summer in the dacha of Skorodumov, and, of course, I spent my summer vacation with them. The dacha was on the shore of the Nevka, and, since the Sul'menevs always had a boat at their disposal, we used to take many trips on the river, going down the Karpovka or sometimes the Chernaia River. The Maresevs lived not far away from us on Peter Island, and we saw them often. Their older daughter Zinaida was then but a few months old.

During my stay at the dacha this time I became gravely ill, and this is how it happened. Once when playing with my cousin Natasha, we decided to water the flowers in the garden. We found two watering cans and went to the barrel standing nearby to fill them. With great difficulty I pulled out the stopper, whereupon the water gushed out with such force that everything around was rapidly flooded. Thoroughly frightened I tried to put the stopper back as fast as possible, but all my efforts were in vain. In my attempt to stop the flow I was soaked to the bone—all the water rushed directly on me. The next day I came down with fever and was bedridden for a long time. Thus I almost paid with my life for a silly prank. I was treated by a doctor who was summoned from the Kamennyi Island hospital. And, thank God, he put me back on my feet before my vacation was over.

I recall also that summer the cannonade from the fortress announcing the birth of Grand Duke Nicholas Nikolaevich. There was an illumination on this occasion in the evening, and that was always very effective on the islands.

So far as I can remember, it was also in 1831 that my mother's friend and her family arrived in Petersburg. She was the widow of General Bublik, who died in Dubno where he was stationed with his division. Her name was Elisaveta Abramovna. She used to send for me frequently, and I recall the very pleasant time I had there in the company of her two charming daughters. Both of them died of scarlet fever. Among the good friends of our family from Radzivilov, General Count Mavriki Gauke also moved to Petersburg. He was appointed to the retinue of the tsar after his brother was killed by the rebels in Warsaw. I spoke of him earlier in my memoirs. I visited him very seldom and therefore do not recall his son who was then a child and who later became famous under the name of *Bossak* during the insurrection of 1863 in Poland.

I felt so at home and happy at the Sul'menevs that I did not like to leave them to visit my other friends and relatives. I used to go with pleasure only to my dear aunt Elisaveta Petrovna in order to see my sisters. Once my aunt gave a ball for children to celebrate her daughter Matilda's birthday. The Institute allowed her to use several large rooms for the occasion, and she invited about twenty of her daughter's best school friends. Several boys and their families, relatives of the girls, were also invited. I, of course, was among them. The ball turned out to be a great success, and never until then had I had such a wonderful time as on that memorable evening. It was the 27th of December, 1831. Many of the girls from the Institute were charming and beautiful. Among them was the slender and graceful Kalinovskaia, who became a lady in waiting on graduating from the Institute. The tsarevich Alexander Nikolaevich fell passionately in love with her. Matters came to such a point, as is known, that she had to be released from the court.

She later married an aide-de-camp, the wealthy Count Branitskii, who, having betrayed Russia and the tsar, moved to Paris. He became our bitterest enemy. He is also known for his friendship with Prince Napoleon (Plon-Plon). However, I liked Eudoxie Butiagin best of all. She was an exceedingly beautiful brunette with fiery eyes. I danced the entire evening with her and we became friendly. For a long time I sighed for her and this, to my great exasperation, amused my sisters who made fun of me. I was then no more than eleven years old!

Among my playmates in the Sul'menevs' home were two very unusual boys—the brothers Echapar. They were Creoles by birth. During his voyage around the world my uncle Fedor Petrovich found them in a very sad state in Manila. He felt sorry for them and brought them with him to Russia.* The elder one, Pedro, who was then fourteen years old, was placed in the Noblemen's Regiment (reorganized later into the Konstantin Cadet Corps), and the younger, Diego, was enrolled in the Naval School. Both were excellent boys, kindhearted and well-bred. They were treated as part of our family. On receiving his officer's commission he (Pedro) was assigned to the Tengin Infantry Regiment in the Caucasus. He took part in several encounters with the mountaineers while serving in the detachment of General Raevskii, who spoke highly of him as an excellent and brave officer. After about three years he asked for a leave and came to Petersburg. He was tired of life in the Caucasus, of the constant marches and skirmishes with the enemy. Tormented by a melancholy premonition, he tried to persuade my uncle Fedor Petrovich to grant him permission not to return to the Caucasus. However, my uncle feared that the kind Pedro with his quick temper would not fare too well

* Their father, Echapar de Breuilles, a native of France, was serving in Spain and held the post of consul in Manila. He was unusually hospitable, especially toward the Russians. Fedor Petrovich was acquainted with him from his previous voyages. When visiting him in Manila in 1828, he found Echapar in great distress. He was in financial difficulties and this worried him greatly because of his large family. With his permission Fedor Petrovich took his elder son Pedro. And Martens, the naturalist in his expedition, took the other son, Diego. When Martens died shortly after their return to Russia, Fedor Petrovich took Diego under his guardianship also.

in peace time under demanding superiors and that he would not be fit at all for such a life. On the other hand, the most brilliant honors awaited him in the Caucasus because of his bravery and because of the constant opportunities for distinction there. Pedro was too devoted to Fedor Petrovich to act against his wishes or even against his advice. He returned to the detachment of Raevskii and was killed in the very first battle! I dearly loved the kind and noble Pedro and sincerely mourned his death.

For some reason I saw much less of his brother Diego. He graduated with honors from the Naval School and became an excellent naval engineer. He married a Skalon and through her became related to the L'vovs. In general, he succeeded in making quite a place for himself in the world. His untimely death followed in 1867. It was a source of profound grief to my uncle Fedor Petrovich who thus outlived both of his foster children.

During the years of my present narrative I met in the home of the Sul'menevs the families Bestuzhev, Opochinin, and Savitskii, who were the Sul'menevs' old friends. The elderly lady Praskovia Mikhailovna Bestuzhev constantly grieved over her three sons, all of whom had been exiled to Siberia for participating in the insurrection of December 14th [1825]. They were all extremely gifted and noble. Among them, Alexander acquired great fame in literature under the pen name of Marlinskii. His beautiful stories enjoyed tremendous popularity. He was permitted to enter military service in the Caucasus. After again being promoted to the rank of officer, he was killed in combat with the mountaineers. His brother Paul, also an extremely gifted young man, was likewise arrested in connection with December 14th, but was reprieved in consideration of his extreme youth. Subsequently, he was an adjutant on the staff of Grand Duke Mikhail Pavlovich. Praskovia Mikhailovna also had several daughters—the oldest, Elena Alexandrovna, already quite a mature spinster, was remarkably intelligent.

Maria Fedorovna Opochinin, the widow of Ivan Savvich's friend in the service, was a bosom friend of Natalia Petrovna. This sensible old lady was unusually talkative, and she amused all of us by her tales which contained a great deal of humor. Of her five sons, the eldest, Alexei, had been living in the Caucasus for a long time. He commanded a regiment there and is now commandant of Tiflis. Vladimir, a retired seaman, was married to the daughter of Admiral Count Geiden, and is known to all Petersburg for his magnificent voice and musical talent. Of the three others, I knew best Alexander Petrovich, who was my brother Alexander's colleague in the service and a great friend of his. As for the family of Savitskii, I remember Valerian Vasil'evich, or, as we called him, Valerushka, a kind but extremely plain young man—a comic figure of awkwardness and naïveté, who was frequently the object of our jokes. His older brother, Nicholas Vasil'evich, a landowner of the Novo-Ladinsk district, was a wise and practical man.

As yet I have said nothing about the oldest and most important member of our family circle, Fedor Ivanovich Engel, an uncle of my mother. He was then president of the Department of the Affairs of the Polish Kingdom in the State Council and enjoyed great influence. The intelligent expression of this little old man, the fine features of his face, his grand manners, his low and fluent speech inspired respect. Strangely enough, he was completely dominated by his wife, who was quite unworthy of him. Her name was Anna Karlovna. She was a tremendous, stout woman and she might have been not bad looking in her youth, but at the time I knew her she was very unattractive in appearance. She had unlimited authority in her household, and everybody feared her. The Engels lived in their own home, which was very elegant, in Kolomna, not far from Tiuremnyi Castle. On Sundays we sometimes gathered there. Everyone was very stiff in their home, which was quite boring to me. However, Fedor Ivanovich was always very gentle and cordial to his relatives. Anna Karlovna's daughter from her first marriage married my uncle Roman Fedorovich Furman, and thanks to her influence the Furmans were closer to the Engels than any other relatives of ours. One of the daughters of Roman Fedorovich, Maria, was even educated in their home. Moreover, on the death of the Engels, who left no children, she inherited practically the entire fortune of Fedor Ivanovich. She married State Councilor Kholodovskii who died soon after the marriage.

Our family circle was expanded at that time by the arrival of my uncles Alexander and Peter Petrovich Litke.

Alexander Petrovich, an extremely striking and handsome man, was married to one of the Furman sisters, Natalia Fedorovna, who loved him passionately and jealously. He was a lieutenant captain in the Black Sea Fleet and arrived from Sevastopol to enter the Ministry of the Navy. The good and kind Alexander Petrovich did not resemble his elder brother Fedor Petrovich either in talents or in education. I must also say the same about my uncle Peter Petrovich, who after serving several years in the mounted artillery left the service to marry a girl of no means or education. I do not recall who her family was. I think her name was Olga Il'inishna. He arrived with her in Petersburg to look for a post in some civil department. He was, however, the kindest man with a congenial disposition. Strange what a difference there was in the education of Grandfather Litke's children of the first and second marriages. In my opinion this is explained by the moral qualities of the two mothers. Ekaterina Andreevna, the second wife of my grandfather, in no way resembled in this respect my intelligent grandmother who, as I said, was Fedor Ivanovich Engel's sister. However, be it as it may, the family relations within the Litke family did not suffer on that account. They all lived amicably with each other, and the Sul'menevs' home was a gathering place for all of them. I believe that it was around this time that Nicholas Petrovich married a sweet, but poor young lady whose family I also do not recall. Her name was Nadezhda Il'inishna.

Early in 1832 the question was solved of how and with whom to fill the junior section of the Imperial Lyceum of Tsarskoe Selo. Up to that time it was formed from the best students from the Lyceum Boarding School connected with the Lyceum. This year was to witness the Lyceum's sixth graduation since its foundation. (Incidentally, my future brother-in-law, Vasilii [Guillaume] Kotsebu, was in the graduating class.) The question led to a number of reforms. At the suggestion of Grand Duke Mikhail Pavlovich, who after the death of Tsarevich Konstantin Pavlovich succeeded him as chief director of the Lyceum, His Imperial Highness ordered that the following measures be adopted.

The number of students was to be doubled from fifty to one hundred. The division into two sections was to be retained, but each section was to have two forms, or classes. Since a complete course in the Lyceum was a six-year course (three years each in the junior and senior sections), each form, of which there would be four, would thus last one year and a half. Until that time all fifty students were maintained at the expense of the state. In order not to increase the Treasury's expenses in doubling the number of Lyceum students, it was decided that the parents of the fifty additional students should pay for their sons' maintenance.

It was impossible actually to carry out this measure in any other way than gradually, because with the sixth graduation in 1832 those who completed the junior section were entering the senior group, or, according to the new plan, the third form. To enroll students in the last, or fourth form, was naturally impossible. Therefore, it was decided to leave the fourth form vacant for the time being. Thus the senior group during the first year and a half consisted only of the third class. In the junior group, or the first and second forms, it was decided that new students would be enrolled. Moreover, it was agreed to form temporarily a preparatory class, consisting of twenty-five students. Thus in a year and a half, after the promotion examinations, everything would be in the proper order. The third form would become the fourth, or the graduating class; the second would become the third, the first would be the second, and the preparatory class would comprise the first form.

On the basis of the reforms, the Lyceum planned to enroll at once in May of 1832 after competitive examinations fifty paying students and twenty-five at government expense. With regard to the latter it was ordered that twenty of the best gymnasium students be selected, five each from the four gymnasia then existing and the other five students from private candidates registered by a special imperial order.

I am describing this reform in detail because it constitutes an epoch in the life of the Lyceum, even if not a very happy one. In his desire to change the old order and to obtain more room, Grand Duke Mikhail Pavlovich ordered discontinued the separate rooms, or, as he called them, the stalls, which served as the students' bedrooms, and their replacement by dormitories. He disliked the secluded life of the Lyceum students, who did not have the right to leave Tsarskoe Selo. He decided to allow them to visit their relatives in Petersburg during vacations, of which there were four a year: Shrovetide, Easter, summer vacation, and Christmas. The students were given three-cornered hats, and the seniors also received swords. Moreover, with the thought of making some sort of civilian pages of the Lyceum students, it was ordered that during the stay of His Imperial Majesty's court in Tsarskoe Selo four students should be sent to the church as guards to the tsar's family on Sundays and holidays. Many of these latter orders were, of course, cheerfully accepted, but whether they had a beneficial effect on the spirit of the Lyceum students is another question.

News of these changes both cheered me and disturbed me. I was happy at the prospect of seeing soon (it was not long before Easter), my brother Alexander whom I had not seen almost since the time I left the Lyceum Boarding School, because we had come to visit him in Tsarskoe Selo only once after that. On the other hand, realizing that only five students would be selected from our gymnasium to be enrolled in the Lyceum, I had little hope of being among this number. Nevertheless, imploring God for help, I plunged ardently into my work to gain this honor and good fortune. Never before had I worked as hard as I did then before the examinations which were to decide my fate. I recall, incidentally, that I read all of Herodotus in Russian translation although it was not required. Finally, the examinations came. I did very well and at the end to my great joy my name was among the five chosen.* My parents' happiness can be imagined when they were notified of this. At that time the Lyceum was regarded as the best school in Russia, and it enjoyed exceptional privileges. My acceptance there, and, moreover, at government expense meant that my entire future was assured. Moreover, I was very happy to rejoin my brother Alexander.

We were ordered to be at the Lyceum after summer vacation, that is, on the first of August. I, of course, spent my vacation with the Sul'menevs who lived then in Pargalov. Alexander and Peter

* To the best of my recollection here are the names of my four schoolmates who were transferred to the Lyceum with me: Khanykov, Murav'ev, Veselovskii, and Galitskii. The first two were placed directly into the second form, and we three into the first. I was the youngest, not quite twelve years old.

Petrovich Litke lived in dachas there too, as did the family of Roman Fedorovich Furman. I became very friendly with one of the sons, Nicholas. The Engels then had their own magnificent dacha on Pargalov Road and we visited them quite frequently. The summer spent in Pargalov left me with the pleasantest recollections. I often took walks in the beautiful park of the owner of Pargalov, Countess Shuvalov, or, after her second marriage, Countess Pot'e.* Moreover, I was completely lost there once and did not return home until evening. I found the entire family of the kind Sul'menevs greatly worried about me.

* She later married a third time, an Italian, Butera, whom she also outlived.

III

THE LYCEUM (1832–1838)

IT WAS WITH GREAT JOY that I rode to Tsarskoe Selo to enter the Lyceum. The weather was beautiful and I was filled with enthusiasm at the sight of the park, which I already knew, and the magnificent palace, whose wing accommodated the Lyceum. Everything seemed to smile upon me, and my future appeared to me in the brightest of colors. We were assigned to our classes, and, although all of us were new, we soon became acquainted and exchanged many questions. I was delighted with my friends with whom I was to spend the next six years almost continuously. From the first day I was aware of a tremendous change for the better in my circumstances, and I pitied the lot of the unfortunate gymnasium students. The administration's polite treatment of us, the wholly decorous environment, both material and spiritual—everything inspired me with a sense of self-respect which I had never felt in the gymnasium.

Our director, Lieutenant-General Fedor Grigor'evich Gol'tgner, lectured to us convincingly, if not eloquently, on how we should behave to preserve the dignity of the Lyceum. Our intelligent inspector, Andrei Fillipovich Obolenskii, constantly talked about the same subject, as did our tutors. Some of them, for example, Chirikov and Kalenich, had been in the institution from its foundation. Anxious to maintain the same noble spirit which always distinguished the Lyceum, they enjoyed telling us about former times. How many interesting stories did we hear from them about Pushkin, Gorchakov, Baron Korf, Del'vig, the unfortunate Kiukhel'beker,[1] and other former Lyceum students who later became famous. However, our doctor, Peshel, distinguished himself more than anyone else in this respect. He was deeply devoted to the Lyceum, and nothing made him happier than the achievements of former Lyceum students in their chosen fields. In his pocket he carried a notebook with the name of every student who had attended the Lyceum from the first year of its existence, with an indication opposite each name of the position occupied by him in the world. It is remarkable, however, that our garrulous Peshel did not like to talk too much about Pushkin: he was hurt by some sharp-witted epigrams at his expense which Pushkin wrote in his school days and which became well known.

An old custom existed in the Lyceum from its foundation, which helped to preserve order and dignity in the institution. The senior group had great authority over the junior group and watched carefully to see that no bad school pranks were indulged in, thus preserving a dignified school atmosphere. The purpose of this supervision, incidentally, was to spare the administration the need of resorting to punishments which, in fact, were practically non-existent in the Lyceum. The students of the junior group always had to behave respectfully toward their elder fellow students and to remove their hats on meeting them. These relations lasted until the final examinations. Once the examinations were over, both groups fraternized and called each other *thou*. These customs had much to recommend them. The freedom which we enjoyed was wisely tempered by rules of decorum and respect for our elders. The latter on their part tried by their good behavior to merit the respect paid them by the younger students. As a result the institution itself gained greatly in dignity and honor. This was the situation in former times. Unfortunately, everything in the Lyceum has changed now, particularly since it was moved to Petersburg.

At the end of August the imperial court moved to Tsarskoe and spent the entire fall there. Our delightful Tsarskoe Selo was greatly enlivened. The Lyceum occupied part of the old palace. We were in the center of all the activities. Before our eyes carriages, dazzling uniforms, and parades in which members of the tsar's family participated were constantly passing. And I must confess that this sight attracted me so much that in the time free from lessons I would not leave the window or the gate of our garden.

We were often visited by our superior, Grand Duke Mikhail Pavlovich, who lived then in Pavlovsk. His stern appearance and military manners often frightened us. He was forever *reprimanding* someone. However, he harmed no one. On the contrary, he had a very kind heart and was always ready to render assistance where needed. Often after scolding all of us, he would joke with us.

Once he entered our class during the German lesson. Calling one of my friends, Sabir, to the blackboard, he ordered him to conjugate the verb *werden*. "Write," he ordered, "the present." Sabir wrote, *Ich bin.* "Fine, and next?" He wrote, *Du bist.* "Wrong," shouted the Grand Duke. Sabir was confused. We started to look at each other, not understanding what was wrong. Our teacher, the kind Petsol'd, puzzled, said nothing. *"Du bin,"* [2] finally said the Grand Duke. "You are a fine one not to know even that," and he burst into loud laughter. Then, on noticing me, I suppose because of my long curly hair (which was still permitted then), he began to ruffle my hair and called me *"shafka* [shaggy dog]." However, he did honor us at times with serious conversation. On the whole he favored us greatly. Sometimes he sent us fruit and sweets from Pavlovsk. He also sent us to the French theater which was open only when the court stayed in Tsarskoe Selo. And on the holidays he would even invite us to parties in his home. This honor, of course, usually fell upon no more than four Lyceum students at one time —two students from each group, or one from each class.

The tsar himself regarded the Lyceum with favor at that time and visited us sometimes. On one such visit he found that the long coats which we wore on weekdays were uncomfortable, and he ordered them replaced by the short jackets which were then worn only by cadets. We did not like this change at all. The tsar was in a good humor on that day and joked a great deal with us and, speaking of the loose tails on our coats, he said that he was replacing them not only because of the greater comfort, but for *economic* reasons as well. When passing our garden during our walks the tsar often talked to us, greeting us in Latin—*quo modo vales!* However, from year to year, for reasons unknown to me, he grew more indifferent and colder toward us and finally ceased his visits almost completely.

On Sundays and holidays four Lyceum students were always selected to act as pages to the tsar's family during the service. The same number of Lyceum students was sent to the court balls, which used to be given for the tsar's children, who were still quite young. The eldest, the presently reigning emperor, was not quite fourteen. Our dancing teacher, Ebergardt, who regarded me as his best pupil, insisted that I did not miss a single one of these balls during my stay at the Lyceum. I cannot express how happy this distinction made me, particularly when I had the honor to dance with the grand duchesses, which happened not infrequently. I recall particularly the balls that used to be given on September 9th, the birthday of Grand Duke Konstantin Nikolaevich.

Around this time (I am referring to the autumn of 1832) my uncle Fedor Petrovich Litke was appointed as tutor to the mentioned grand duke. This occurred, as far as I know, in the following manner. Soon after Fedor Petrovich was appointed aide-de-camp, the tsar told him that he intended to intrust him with the education of his second son, Grand Duke Konstantin. The latter was granted the rank of grand admiral at his birth and was intended for a naval career. My uncle was greatly disturbed by this decision which in no way suited his nature and his plans for the future. He was then engaged in writing a descriptive account of his last voyage around the world. Moreover, on the advice of Humboldt and other scientists with whom he was in constant correspondence, he was thinking of other ocean voyages in order to complete his discoveries and observations. To abandon his cherished plan, to give up his chosen field, to which he was passionately devoted and in which he had already distinguished himself, and to launch upon the education of a little six-year-old boy, to be constantly with him, and to teach him his ABC's—this seemed intolerable to Fedor Petrovich. Moreover, he did not regard himself as suited to such a post by nature and training. He told the tsar all this frankly, imploring His Highness to choose another tutor for the grand duke and one more suited for this work. The tsar listened patiently to his arguments and was persuaded to grant his wish. After consultation, the tsar decided to appoint Lieutenant Captain Vukotich as tutor to the grand duke. Vukotich, if I am not mistaken, at that time held the post of either inspector or company commander in the Naval Cadet Corps. My uncle was triumphant, but not for long. Hardly was Vukotich appointed to the post than he became sick and died. The tsar then summoned Litke and told him this time quite emphatically: "It seems that it is God's wish that you be my son's tutor and so do not decline!" With a heavy heart Fedor Petrovich accepted the tutor's responsibility which proved to be by no means an easy one with a gifted, but extremely willful little boy.

Since he had to be with the grand duke from that time on, Fedor Petrovich used to move with the court to Tsarskoe Selo in the spring and autumn months. It was then that my brother Alexander and I used to spend our holidays in his home, or, to be more exact, in the court quarters which he occupied, because, unable to leave his charge, he would go home but once or twice a day for a moment when he had to change his clothes. Nevertheless, we were always happy to visit him because of the full freedom he gave us, and also the delicious food which was served to us from the court kitchen. We were served the best wines and a remarkably delicious mead (Hydromel), which we relished. I also liked the fact that the young grand dukes lived in the same wing of the Alexander Palace where the apartment of Fedor Petrovich was. I met them often in the hall and became acquainted with them. All activities in the palace interested me tremendously. In the mornings, the trooping of the colors, then the arrival for dinner, the military music in front of the palace, and finally the evening ceremonies during holy days. All this passed before my eyes and all this thrilled me. My brother Alexander was of a different nature. He was satisfied to sit quietly all day in my uncle's study reading a book. In my defense I must say, however, that my brother was five years older than I.

With the departure of the imperial family the usual quiet descended upon Tsarskoe Selo. Our school work went on as usual. My life with my good friends was infinitely more pleasant than it had been in the gymnasium. I had become particularly friendly with the intelligent, caustic, but quite lazy, Ginter, who had formerly been educated in Nezhin. He told me many interesting facts about the life and customs of Little Russia. I was also friendly with the brothers Voeikov, with Veselovskii, who was always the best student, and others. I can honestly say that I never quarreled with any of my friends. Among them I would like to mention Sabir, Galitskii, Baron Ikskul, Gernet, Kaidanov, Obukhov, De Brin'i (De Brigny), Bogaevskii, Akhsharumov, Marchenko, Petrashevskii (known for the unfortunate incident for which he was exiled to Siberia),[3] Zakharzhevskii, Stessel', Trofimov, Teplov,

Nekliudov, Arsen'ev, Iakhontov, and Kreiton. The son of our director, Nicholas Gol'tgner, was also in our class. He was preparing
for a military career and wore the page's uniform. There were
twenty-five of us in the class. Only twenty-two remained on graduation, and at the present time there are no more than six or seven
of them still alive. Our class is regarded as one of the most unfortunate since the foundation of the Lyceum.

At first I used to spend Sundays at the home of our professor of
law, Baron Egor Vasil'evich Vrangel, who had a very large family.
His older son, Vasilii, was my brother's friend and was at the top
of his class. After the family of my friends, the Voeikovs, moved
to Tsarskoe, I used to visit them more frequently than the
Vrangels. There I was treated as a member of the family. Their
father was no longer living. Their mother, Vera Nikolaevna, born
L'vov, was Derzhavin's [4] granddaughter and the sister of the then
aide-de-camp, Colonel Alexei Nikolaevich L'vov, the famous
musician and composer of our national anthem, *Bozhe Tsaria
Khrani!* [God Save the Tsar!] In order to be nearer her sons, she
lived in Tsarskoe Selo with her daughter Maria (later married to
Polenov), a very sweet young lady who had many admirers among
the Lyceum students.

Thus I was not lonely in Tsarskoe. Nevertheless, I awaited impatiently the arrival of the holidays which gave me the opportunity
to go to Petersburg and spend two weeks there with the kind
Sul'menevs. On the day set for our departure, our old friend
Evstafii would arrive with the carriage in which we would ride for
twenty-five versts over the smooth sleigh road.

Our arrival during the holidays was always a joyous event at the
home of the Sul'menevs, where we thoroughly enjoyed ourselves.
Everything was simple in their home. The expensive wax candles
were lit only on holidays (the stearin candles, so common now,
were not even heard of then), the ladies' dresses were simple, not
like now. Then, thank God!, unnecessary and ruinous pomp was
not in our [Russian] nature. Our pleasures were not poisoned by
the bankruptcy to which the luxury inevitably leads that is spread-
ing among all classes of society. When people yield to their irra-
tional demands they loose their peace of mind and deprive themselves of genuine quiet pleasures, and even of simple kindnesses,

such as hospitality, which is almost impossible for a man under the present conditions of life unless he is rich.

However, let us return to Tsarskoe Selo. I regard the six years in the Lyceum as the best years of my life, which, of course, speaks well for the institution. I cannot say that I learned there everything that a well educated young man should know. Nevertheless, I did learn there the usefulness of knowledge, and I acquired a desire to learn. And, what is most important, I acquired the characteristics of honor and nobility of mind which were at the heart of the Lyceum education. I am not saying that in our time there were no exceptions in this respect among the Lyceum students, but those exceptions were rare, and we had practically nothing to do with such persons.

As I have already said, I was not quite twelve years old when I entered the Lyceum. I was the youngest in my class. Some of my classmates were almost three years older than I. In spite of this difference in age, I was at the top of the class in the first years; but I began to lag considerably in the upper classes so that in the last year I stood ninth in my class—largely because I found ethical philosophy and higher mathematics very difficult and also because I abandoned myself to pleasures and amusements during the stay of the imperial court in Tsarskoe. I was always more studious during the winter months.

On my promotion to the next class, that is, in January of 1834, the son of Vice-Admiral Golovnin, mentioned earlier in my memoirs, entered the Lyceum. After the death of this famous navigator of ours, his widow, Avdot'ia Vasil'evna, born Lutiaevskii, devoted her life to her children, from whom she never parted. When her son entered the Lyceum, she moved to Tsarskoe where she lived for the next six years. I had met the Golovnins in Petersburg before through the Sul'menevs, who were distant relatives. Avdot'ia Stepanovna [sic], a woman of remarkably fine mind and education, was very attentive and kind to me, and I used to visit her home gladly during the holidays although there were no particularly gay entertainments and everything was reserved and quiet. Her four daughters would invariably be busy either with fancy needle work or reading. The poor girls knew nothing of dances, the theater, or other worldly pleasures. One of them married Salomon, a friend of their brother Alexander Vasil'evich. However, the others have remained spinisters and continue their monotonous life with their aged mother who must be around eighty years old.

The year of Golovnin's graduation marked the most fortunate and successful year of the Lyceum since its foundation. Alexander Vasil'evich himself became minister of public education. Practically all his classmates distinguished themselves. Among them, Reitern [Reutern] became minister of finance and Baron Nikolai head of the civil administration in the Caucasus. I used to meet these gentlemen at the home of the Golovnins, and they became my good friends. Alexander Vasil'evich seemed to me to be the most intelligent among them. He exerted a great influence over his friends. Reitern had good common sense, poise, and a fine sense of values, and Baron Nikolai was remarkably gifted and witty. He grasped everything very easily.

Our graduation was the ninth from the time of the foundation of the Lyceum. Golovnin, Reitern, and the others belonged to the tenth graduation. The next graduation year, the eleventh, whose students entered the Lyceum three years after me, also gave us two ministers, Count D. Tolstoi, minister of public education, and Count Alexei Bobrinskii, minister of public works. Since they were considerably younger than I, we did not see much of each other.

Our unfortunate graduation year, the ninth, produced few prominent men, except for Konstantin Stepanovich Veselovskii, who chose a scientific career and became permanent secretary of the Academy of Science, Obukhov, who for a time was acting minister of interior, and perhaps myself, advancing in the diplomatic sphere, also Arsen'ev who died with the rank of governor of Tula. None of my friends went too far. Nekliudov, a very gifted man, chose the diplomatic career and was promoted much more rapidly than I. He was soon appointed chief secretary of the mission in Athens, but his career was cut short because of the rashness and the instability of his character. Kreiton was very promising, but, unfortunately, he died shortly before graduation. Baron Ikskul remained in government service but a short time and then left it to manage his estate in Estonia. The same can be said about Galitskii, a wealthy Saratov landowner, who was in the service only about a year or two.

Kreiton's father, a famous physician, was private doctor to

Empress Alexandra Fedorovna and followed Her Majesty wherever she went. During the court's stay in Tsarskoe Selo he occupied an apartment with his large family in the Alexander Palace, on the second floor, over the arsenal. I visited him frequently and was always very hospitably received in his home because of my friendship with his son. Dr. Kreiton's wife, born Sutgaf, was a well educated and amiable woman. She liked to see young people enjoy themselves in her home. In the evenings we either danced or played charades. Among the other Lyceum students who visited the Kreitons were Shtorkh (a friend of my brother), Naryshkin, and Baron Nikolai. The latter was a distant relative of Mrs. Kreiton. Her brother, General Sutgaf, was married to Baron Nikolai's sister. We spoke only French in this company. I recall how we rehearsed and presented a play by Scribe, *Le cuisinier et le secrétaire*. We gave it at Christmas time in Petersburg at the home of Mrs. Kreiton's father, the elderly Sutgaf, who, if I am not mistaken, was the director of a boarding school. Shtorkh (he is now secretary of state and head of the IV Department of His Majesty's Own Chancery) took the part of the cook and I was the secretary. I was barely fourteen years old then, but I was tall for my age and, therefore, was chosen for the part.

In those years I grew so rapidly that each time I appeared in Petersburg during the holidays my relatives would greet me with shouts of astonishment. I outgrew all classmates. In some respects this was not to my advantage because when the tsar or the grand duke visited the Lyceum and we were all lined up, I always had to stand first and, consequently, was the most conspicuous.

Speaking of my friendly relations with the Voeikov family, I cannot help but recall that in 1833, or in 1834, while working on the composition of our national anthem, *Bozhe Tsaria Khrani*, Voeikov's brother, aide-de-camp Alexei L'vov, who at the time was attached to the imperial court in Tsarskoe Selo, once decided to try out the anthem. For this purpose he organized a small choir at the home of the Voeikovs in which took part his younger sisters, who came from Petersburg and who were not yet married, Maria Voeikov, Borozdina, his niece,* two or three men whose names

* She later married a friend of mine, Leonid Voeikov (her cousin), and died shortly afterward.

I do not recall, and myself. Both in the Lyceum and the gymnasium
I was always among the singers, and frequently I was called upon
to sing in the Russian Greek Orthodox Church although I was a
Lutheran. After several rehearsals under the direction of L'vov
himself, we went to the so-called rotunda, a round building with
a dome in the center of the Chinese village and consisting of only
one large room. The rehearsal was a success. No one in the audience
had ever before heard the anthem which shortly afterward replaced
the former English, "God Save the King," and everyone was excited. Whenever I hear the sounds of our national anthem I recall
this incident with particular pleasure.

The summer of 1833 and 1834 found the Sul'menevs living somewhere in Okhta, near the Kushelev-Besborodok garden. My brother
and I used to go there during our vacations. I recall that during the
summer of one of these years it rained so much that it was only
very seldom that we were able to go out for a walk. I have never
seen anything like it since. From boredom we used to spend whole
days playing cards with the elderly Sul'menev who frequently reprimanded us for our poor game.

As far as I can recall, it was at the end of 1834 that Katin'ka
and Anneta Sul'menev graduated from the Patriotic Institute.
Anneta was unusually lively and beautiful. Her older sister,
Nadin'ka, who was being educated at home, was perhaps even more
beautiful because of her more regular features. The gentle and intelligent Katin'ka was very friendly with us. Natasha was still very
young. At about this time also their older brother Nicholas graduated from the Naval Officers' Corps. He was jolly and witty, but
somewhat willful. He amused us to tears with his funny and
original tales and jokes. The presence of all these young people
filled the house with life. The hospitable Sul'menevs entertained
many people, and on some days of the week we used to dance till
morning.

In January of 1835 fate brought me to their home in a most unexpected and pleasant manner, and not during vacation time. And
this is how it happened. A children's masked ball (*bal costumé*)
was given in the Winter Palace for the grand dukes and duchesses
who were then very young. Our director received instructions to
select four Lyceum students, one from each class, for this ball. In

such circumstances, as I have already said, they never overlooked me, and I was among the ones chosen. General Gol'tgner took us to Petersburg and we stayed at the house on Haymarket Square reserved for the Lyceum students who were to be presented to Grand Duke Mikhail Pavlovich, who frequently invited us to the children's balls. At the appointed hour we arrived with our director at the ball. I cannot describe my admiration of the magnificently illuminated ballroom of the Winter Palace. The tsar received us most graciously and took part in the dances himself. Except for the pages and Lyceum students, all the children were in masquerade costumes. The grand dukes, of course, were also in regular clothes. The heir wore a Cossack blue coat (he did not wear an officer's uniform yet) and Grand Duke Konstantin Nikolaevich wore a naval jacket. After one of the dances the tsar came out into the middle of the ballroom and ordered everyone to stand back. When this was done and everyone was silent, the door opened at a signal from the tsar and a platoon of armed cardboard soldiers, marching in good order, entered the ballroom to the tune of an old march. The tsar took command of the troops and, overcome with laughter, began to give various orders which the men executed perfectly. It turned out that young cadets were concealed in these cardboard figures. After this act, which was a great success and most amusing, the dancing was resumed.

In this connection I recall one incident which almost caused me some trouble. During the ball my uncle Fedor Petrovich, whom I was so glad to meet there, came and said he would like to introduce me to a young lady who would like to dance with me. I do not know whether he mentioned her name. At any rate I did not hear it. After bowing to the charming young lady, who was elegantly and richly dressed, I took her hand. She pointed to the places where we were supposed to stand. It was a quadrille with all the grand dukes and grand duchesses participating. Somehow I did not pay particular attention to this. On engaging my partner in conversation, I discovered she was intelligent and gracious and, most important, she seemed to know all those present at the ball and very willingly satisfied my curiosity when I asked her about anyone. I was so pleased with her that I asked her there and then if she wouldn't dance the next dance with me. She willingly accepted my invitation for one more quadrille and a mazurka. Thus I danced with her practically the entire evening, unconcerned as to who she was and thinking that I could find that out after the dance. For the time being I was intent on the dance, particularly since I was constantly called upon to take part in various steps, and always in the circle of the tsar's family. I was carried away by the dancing and in my childish artlessness passed on my enthusiasm to my partner. However, from time to time the thought did cross my mind about why I had the honor of dancing with their highnesses and why our strict headmaster Grand Duke Mikhail Pavlovich was constantly in back of me and why was he joking at my expense? Finally we were called to supper. Uncle Fedor Petrovich had told me at the beginning of the ball that the Sul'menevs were having a large gathering with dances that same evening. Wishing to take me there with him, he approached General Gol'tgner to ask his permission for a twenty-four-hour leave for me in Petersburg. The director gave his consent and added that Giers had conducted himself excellently at the ball and had the honor of dancing with the grand duchesses. On taking leave of Gol'tgner I asked my uncle who the charming lady was to whom he had introduced me and with whom I danced so much. "But didn't you know that it was Grand Duchess Elisaveta Mikhailovna?" [the grand duke's daughter] answered my uncle. I confess that I was paralyzed with fright. No one had the right to invite the immediate members of the tsar's family to dance. This strict and widely known regulation was frequently impressed upon us. Had the director known that I had violated this rule, even though unwittingly, he would have never again chosen me to attend the tsar's balls even if he did not punish me. And the grand duke, I thought, apparently suspected my mistake, and, in spite of it, he had been so kind to me! Be it as it may, I exercised caution and for a long time told no one in the Lyceum about this *faux pas*.

And thus I finished the evening in the home of the Sul'menevs where my appearance with my uncle Litke at 2 o'clock in the morning was a happy surprise to all. There I danced until morning. After spending the rest of my twenty-four hour leave with the Sul'menevs, I returned to Tsarskoe Selo.

I spent the Holy Week of 1835 in the home of the Sul'menevs, and it was there that I received an order to appear before Grand Duke Mikhail Pavlovich to attend the sunrise services and the breaking of the fast. Practically no one remained in the Sul'menevs' house. Everyone was in church. At midnight I dressed and went out in the streets hoping to find a cab, but, unfortunately, I did not meet a single one and had to proceed on foot to the Mikhail Palace. The distance from the Eighth Line on Vasilii Island is so great that I could easily have been late. Consequently, I no longer walked, but actually ran. This was not easy because a blizzard began although it was spring. I was greatly worried, but finally, out of breath, I reached the palace. I hardly had time to shake the snow off my coat when the grand duke arrived from the Winter Palace where he had attended the sunrise services. He exchanged the "Christ has risen" greetings with us and gave each of us a china egg. We were then invited to the luxuriously arrayed Easter table. The Grand Duchess Elena Pavlovna was at that time bewitchingly beautiful. Everyone admired her magnificent stately appearance, her wise and graceful manner with all the guests. Among them Adjutant General Paul Dmitrievich Kiselev [5] impressed me particularly both because of his remarkably handsome appearance as well as because of his fine and noble manners. He had returned not long before that from the Danubian Principalities where he made himself famous by his wise administration.

On the eve of our examinations my brother and I received the sad news of the death of our father, which occurred in Radzivilov on April 20, 1835. I recall that when walking in our Lyceum garden to my surprise I suddenly saw my uncle Ivan Savvich leaving the apartment of the director. Overjoyed I ran to meet him when he entered the garden. I sent for my brother at once. While walking with us in the garden our kind uncle mentioned Radzivilov and said that he had news from there that was not good. My brother and I were silent with a foreboding of some misfortune. Your father is very sick, added Ivan Savvich and with tears in his eyes he embraced us. I cannot express the profound grief I then felt. I understood that our most kind and tenderly beloved father was no longer living! After sharing our tears with us and adding his blessing with genuine Christian feeling, our uncle told us that we should no doubt see our mother as soon as she was ready to go to Petersburg.

Here is what I later found out from my mother about the last days of our remarkable father's life. His illness, the only symptom of which was a loss of strength, lasted but a few months. This exhaustion began after the doctor healed an open wound on his foot. Up to that time he had enjoyed excellent health and was practically never sick. However, now he began to loose his strength and frequently felt faint. As if sensing that his end would soon come, my father often spoke of his absent children and wondered how he might be nearer to them. On the advice and pleadings of my mother he filed a petition to be transferred from Radzivilov to which he was so greatly attached. Because of his good connections and excellent reputation, this presented no special difficulties and his appointment to the post of head of the Riga District Customs was almost decided upon when his premature end came in the fifty-eighth year of his life. His death brought grief to all those who knew and respected his noble character and kind heart. He was buried in Dubno and according to the Russian Greek Orthodox rites, although he was a Lutheran. A tremendous number of people came to attend the burial. The local archbishop himself officiated and delivered a funeral oration which subsequently appeared in print. I wanted to preserve a copy of this sermon for myself, but to my great regret I can no longer find it among my papers.

On learning of my father's death, his older brother, Fedor Karlovich, rushed from Kamenets to Radzivilov. The two brothers had been bound together by the closest friendship. In the spirit of genuine brotherly affection, of which, unfortunately, there are so few examples in these days, he took under his wing the family of his mourned brother and friend. Fedor Karlovich helped mother settle her family affairs and offered to move her with her children to his home in Kamenets. My mother accepted this offer on the condition, however, that she could first go to Petersburg to see us, but chiefly to call for and take with her my sisters Emilia and Valeria who were graduating from the Institute at that time. My most kind dear uncle Fedor Karlovich accompanied my mother to Petersburg. They took with them my eleven-year-old brother Fedor

to place him in some school. And my younger sisters Iulia and
Anneta were left in Dubno with the Sitnikovs.

In the meantime our examinations at the Lyceum began. My
brother Alexander who had always been known for his ability and
diligence passed his examinations brilliantly and graduated in the
ninth rank (titular counselor) with a first silver medal. I also did
fairly well in my examinations, and on being promoted to the next
form, I was sixth or seventh in my class. Of all the subjects I feared
Latin especially. I did poorly in this subject and for this reason was
naturally not liked by the instructor of this language, Felagont
Vasil'evich Grizdov. The other subject in which I did not do well
was Catechism because as a Lutheran I was taught this subject in
German, which I spoke with great difficulty. Because of my
Lutheran faith everybody assumed that I was German, and people
were surprised that I expressed myself so badly in my own native
tongue. This both embarrassed and angered me. Aside from religion I had nothing in common with the Germans, and I regarded
myself then, as I do now, as a pure Russian although my ancestors
were Swedes. In this connection I cannot refrain from telling of a
circumstance which puzzled me greatly at the time. While preparing for the examination in Catechism, I spent practically the entire night over a book and went to bed only toward morning. For
a long time I was unable to fall asleep from worry over the forth-
coming examination. Finally I fell asleep, and in my dreams I saw
our pastor Avenarius, with a mysterious and inspired look on his
face, entering the church with two tiny coffins under his arms. I
was so awed by this dream that I awoke. At 8 o'clock in the morning we were all in the examination auditorium. The Reverend
Kochetov was already there. We were waiting for the pastor
Avenarius. Suddenly our director entered and said: "You will have
no examination today. A great misfortune befell your poor pastor.
He lost two infant sons tonight. They died of scarlet fever." I shall
not attempt to explain this remarkable incident, but I do vouch for
its authenticity.

At the end of May our mother came to call for us from Petersburg
where she had been for several days awaiting the end of our examinations. I cannot express my emotion on meeting my mother
whom I had not seen for more than six years. With tears in my

eyes I threw myself on her neck, and we were inseparable from that moment on. I had felt this strong love and attachment toward my mother since early childhood. And these sentiments never lessened despite the long separation.

My mother arrived in Tsarskoe Selo together with my dear uncle Fedor Karlovich and kind Ivan Savvich, and also my younger brother Fedia, whom, of course, I would not have recognized. He was a jolly and bright little boy. He amused all of us by telling us in his broken Russian, mixed with Polish, of all the wonders he had seen on the trip. He seemed just as thrilled as I had been on my first arrival in Petersburg, seeing everything in a distorted fashion.

We left in two carriages and drove over the magnificent Petersburg road straight to Novaia Derevnia, to the dacha of the Sul'menevs. My mother stayed there during her visit in Petersburg together with my sisters Emilia and Valeria, who had just graduated from the Institute, also my brother Fedia. My brother Alexander and I remained here too. Alexander was preparing to enter the service. Naturally, I remained here during my summer vacation. Almost no one now with the present mode of life could understand how two such large families could be housed in one dacha. But we were all very comfortable and happy, and contentment, cordiality, and genuine family harmony never ceased to reign in our midst.

Except for my two younger sisters Iulia and Anneta, who, as I have already said, had been left in Dubno with my brother-in-law Sitnikov, our entire family was here with the kind Sul'menevs.

Uncle Fedor Karlovich was staying with his brother Alexander Karlovich on Vasilii Island, but he visited us practically every day. He was of a cheerful disposition, loved to tell stories, and always spoke with great enthusiasm and interest; he frequently exaggerated, but this weakness was redeemed by his pleasant courtesy, consideration, and fine manners. He was then about sixty years old. His only daughter Valeria was placed, when her mother died, in the care of our dear aunt Anna Karlovich Maiet and was with her in Kamenets Podol'sk at the time.

Soon after our arrival in Novaia Derevnia, the emperor's court moved for a time to Elagin Island. Thanks to its proximity, Uncle Fedor Petrovich Litke was able to visit us often. He was at that time promoted to rear admiral with an appointment to the retinue of His Majesty. We all rejoiced over this rapid promotion. I recall that on the occasion of one of his visits he announced that he was about to marry a young lady, Miss Brown, the tutor of the beautiful Grand Duchess Alexandra Nikolaevna. Several days later his fiancee came with her mother to meet her bridegroom's relatives. She made the pleasantest impression on all of us. Julie Brown,* or Iulia Vasil'evna as we called her according to the Russian custom, was really a charming young lady. She was striking and rather beautiful. Intelligent, well educated, and of fine character, she was a worthy companion for a man as remarkable in every respect as Uncle Fedor Petrovich.

On June 29, the day of SS. Peter and Paul, all our relatives gathered at the Sul'menevs to celebrate their silver wedding. My poor mother, whose happy married life was cut short by the will of God only two months before this date, also a memorable one for her, was sad the whole day. I believe I have already mentioned the circumstances under which the weddings of my parents and the Sul'menevs took place on the same day in Radzivilov. Deeply religious and strong of spirit, my mother bore her misfortune with Christian resignation and, although she never ceased to mourn the loss of my incomparable father and wore mourning clothes to her death, she nevertheless always fulfilled her maternal duties with love toward her children.

Not wanting to deny us, despite our mourning, the pleasure of enjoying the Peterhof celebration which took place on July 1, Empress Alexandra Fedorovna's name day, my mother went with us to Peterhof. We stayed in one of the court houses, near the palace, which was assigned to Uncle Fedor Petrovich. The entire family of the Sul'menevs was also with us. Unable to leave his august charge for any length of time, Fedor Petrovich himself came home but seldom, and we took possession of his quarters. Although we were numerous, as his guests all of our food came from the tsar's kitchen where Fedor Petrovich was entitled to order what he pleased. Only in the Russian court are such luxuries and lavishness possible! On

---

* She was no relation to the young ladies of the same name who lived with the Sul'menevs.

the day of the celebration Fedor Petrovich dined with us. There were twenty of us at the table, and the dinner was most enjoyable. I see before me, as if today, my exuberant Uncle Fedor Karlovich telling the most remarkable stories with great enthusiasm. In the evening we went to see the illumination. It is impossible to imagine anything more beautiful and elegant than the Peterhof garden illuminated by thousands upon thousands of lights, reflected in the sea, in the ponds, and in the fountains. Particularly remarkable was the square in front of the palace with an enormous fountain and a statue of Samson. Music came from all parts of the garden, and there were so many people that one could barely proceed along the garden paths. The most interesting moment was when all the tsar's family, accompanied by the members of the court and surrounded by cavalry guards and various courtiers on horseback, drove through the garden in the elegant narrow carriages which for some reason one never sees any more. Here you took your life in your hands if you tried to go closer and enjoy this picture. Such celebrations no longer exist, and it is difficult to imagine them from descriptions, no matter how detailed. One simply had to see them for himself. Such at least is the impression they made on me.

On returning to Novaia Derevnia we often visited the Engel'-gardts who lived on Aptekarskii Island near Laval's garden in Seliavin's dacha. This was my first introduction to this family with whom I later spent several years. Paul Vasil'evich Engel'gardt, a retired colonel of His Majesty's Own Uhlan Regiment, was a very rich man. His father, thanks to his close relation to the famous Prince Potemkin Tavricheskii,[6] from whom he inherited an enormous fortune, was a full general at the age of twenty-two. Paul Vasil'evich and his elder brother Vasilii Vasil'evich left the military service with the rank of colonel and divided their father's inheritance between them. Vasilii Vasil'evich who, incidentally, owned one of the most magnificent homes in Petersburg on the Nevskii Prospect, where the Noblemen's Club was housed for a long time, lived too lavishly and soon lost his fortune. Paul Vasil'evich was different. He was more than prudent, and although the furnishings of his home revealed him a rich man, he lived very quietly and quite differently from his brother who indulged in sumptuous feasts daily. This mode of living suited perfectly the taste and character of his wife, Sofia Grigor'evna, a charming and unusually beautiful woman, who was very shy and did not like worldly pleasures. Sofia Grigor'evna was also born Engel'gardt, although not related in any way to her husband.

About three years before the death of my father, Paul Vasil'evich stayed with my parents on his return from abroad with his family. My parents received them with their usual kind hospitality. The two families became great friends and Sofia Grigor'evna, in particular, grew very fond of my kind and intelligent mother. The Engel'gardts took a sincere and close interest in my mother's grief when she had the misfortune of losing my father, and they visited her frequently after her arrival in Petersburg. On their insistent invitation, my brother Alexander, who was looking for a post after graduating from the Lyceum, came to stay with them and became, one could say, a member of their family. Our close relations with the Engel'gardts date from that time.

Since my mother and practically all our family were in Novaia Derevnia I spent my time there most pleasantly. There were no railroads then, and very few Russians thought of trips abroad, which would have been difficult and expensive. The excellent music of the cavalry guard regiment, the most brilliant in the empire, attracted many people to Novaia Derevnia, which had an added attraction—the Institution of Synthetic Mineral Waters where magnificent entertainments and balls were given. The master of ceremonies of these entertainments was the well-known magnate Prince Sergei Gagarin. We, of course, did not participate in these entertainments directly because of our mourning, and yet they greatly affected our life in the dacha.

On August 1 my summer vacation ended and I had to go back to Tsarskoe Selo. This time I was sad at the thought of my return, particularly since my mother was preparing for the trip to Kamenets with my sisters where she planned to remain in the home of my kind uncle Fedor Karlovich. He, on his part, was in a hurry to return home. "Beautiful as Petersburg is," he frequently used to say, "my home in Kamenets is better; I have so much more room there." And he often sighed over the wonderful climate and beautiful scenery of Podolia. He departed from Petersburg early in August, and my mother followed him in September.

On my return to the Lyceum I proudly placed the gold insignia on the collar of my uniform coat in place of the silver ones that I had worn in the lower form. I was also pleased with my new quarters on the second floor which had much more room than those on the floor where I had spent the first three years. I also felt proud to be a real student. In the upper forms our instructors were professors and not assistants or ordinary teachers as in the lower forms. Moreover, our tutors were different too, and they allowed us more freedom than before. The kind and vigilant Fotii Petrovich Kalenich, who took his duties of pedagogue so seriously, was busy with a new batch of freshmen, or little birds as he called them, instilling in them the spirit of the Lyceum with which he himself was profoundly imbued as one who had been associated with the Lyceum from its foundation. I must stop and say a few words about this unique and admirable person.

Fotii Petrovich began his career in the choir of the court chapel. Being capable of excellent handwriting and having gained the well-deserved reputation of being an honest and highly moral man, he was appointed instructor in penmanship and a supervisor in the Lyceum. He thoroughly understood his duties, which he considered a lofty calling, and executed them with love. He was strict, but he never resorted to punishments. Instead he argued with the culprit to convince him of his guilt and to make him repent. Possessing neither a great mind nor education, the kind and firm Fotii Petrovich spoke the language of the heart in his work of rearing young boys. This was most important. To be sure, he did talk too much and often talked himself into absurdity and was completely incomprehensible in his attempt to be profound. But we listened, nevertheless, to this venerable old man and sincerely respected and loved him. We tried not to displease him, and we were on our best behavior while he was on duty.

On one occasion we had Adjutant-General Ivan Onufr'evich Sukhozanet, temporarily acting chief director of the Lyceum in the absence of Grand Duke Mikhail Pavlovich. The crippled Sukhozanet (he lost one leg in the Polish campaign) was a great eccentric. He liked to deliver orations and, like Kalenich, indulged in endless philosophizing. Once, after making the rounds of the classrooms and talking to his heart's content, he went into the administration office and announced to the director that he would like to test personally the ability of our tutors and to find out how they understood their duties. He ordered that they all be assembled in the administration office for this purpose. Here the examination began. None of the tutors, not even Andrei Fillipovich Obolenskii, who had the reputation of being the most intelligent, could satisfy Sukhozanet with his replies, until he addressed Kalenich. "What is a tutor?" he asked. Fotti Petrovich was not embarrassed in the slightest. He rose to his feet, assumed the pose he invariably took when about to reprimand us, and, after taking a pinch of tobacco from his box and waiting a second, he waved his hand and with gravity replied: "A tutor? That is the mirror in which the student must find his own reflection." Stunned by this unexpected reply, Sukhozanet was enraptured, showered Fotii Petrovich with a thousand compliments, and held him up as the model to the rest of his puzzled colleagues.

On another occasion (in 1834) Sukhozanet * had the idea of making all of us report to the Lyceum in the middle of our summer vacation. Frightened, we assembled in Tsarskoe Selo, none of us understanding why we were ordered to come. Most of us, including our superiors, thought that the institution was about to be closed, just as several years before the Lyceum Boarding School had been dissolved. Finally Sukhozanet appeared and ordered us to line up in the large auditorium of the senior course. He gave us a long speech, I do not know on what subject, but we couldn't make any sense out of it, and, finally, raising his voice, he said that during the celebration in Peterhof he had met a Lyceum student wearing a brimmed hat (contrary to regulations) and that when seen he had disappeared down one of the lanes. "Which one of you was it?" he asked. Baranov, a friend of my brother, stepped out and said, "It was I." Sukhozanet came up to him and praised him for his honest confession. He then congratulated us for having in our midst such an excellent school friend and proceeded to deliver another speech, with many references to Russian history. Finally he dismissed us. It should be mentioned that although Baranov is

* His brother, Nicholas Onufr'evich, held the post of minister of war following the Crimean War.

really a most upright man, he also happened to be a nephew of Count Vladimir Fedorovich Adlerberg.[7] Perhaps, even if the assumption should not be made, Sukhozanet would have acted differently had the culprit been someone else among us. Be it as it may, we found Sukhozanet quite a bore and were happy when Grand Duke Mikhail Pavlovich returned.

Before departing from Petersburg my mother visited me twice in Tsarskoe Selo. I recall that on one of her visits we spent an evening in the home of Iulia Vasil'evna, the bride of Uncle Fedor Petrovich Litke. Their wedding took place several months later in Petersburg, but I was unable to attend.

If I recall correctly, it was also in this year that the Alexander column was erected on the Palace Square. To celebrate the occasion, the Lyceum students were called to Petersburg on Alexander Day [name day of Alexander], August 30. We assembled in the halls of the Pages Corps, and from there we were marched to the square. After the ceremony we were dismissed for the day, and I went to the Sul'menevs' home.*

Due to a petition addressed to Prince Peter Oldenburg by my uncle Fedor Petrovich, my younger brother Fedia was placed on the list of candidates for the Imperial School of Jurisprudence, a privileged institution, which was then being founded and where he was subsequently accepted and educated at the expense of the government. For the time being, before my mother's departure from Petersburg, he was placed in the pension of Shubert on Vasilii Island.

Finally the day of my separation from my mother came. She traveled by way of Tsarskoe Selo where she spent a whole day with me. I cannot express my profound grief at this separation. I seemed to have had a presentiment that this separation was to be forever. My mother departed with my sisters Emilia and Valeria. For a long time I was inconsolable, but the daily routine was too engrossing and I continued my life at the Lyceum as before and felt fairly happy.

* On looking up the dates after writing these lines, I discovered that the unveiling of the Alexander column which we witnessed took place not in 1835 but in 1834.

As I have already said, I was not doing as well in the upper forms as I did in my first years at the Lyceum. I was not mature enough for the higher studies. Nevertheless, I managed to follow the others and was seventh or eighth in my class. In some subjects, more within my reach, such as history, political economy, and statistics, I was even one of the top students. Among the professors whom I recall most pleasantly were Baron Vrangel' (Roman Law), Kaidanov (History), Ivanovskii (Political Economy), Shul'gin (Statistics), Zhile (French Literature), and even the professor of German literature, Oliva, although I did not do too well in his course. Oliva was a remarkably intelligent man and an effusive one. He read with great animation and, in spite of his ugliness (he had no nose), his facial expression was very pleasant when he talked. We disliked our inspector and professor of Philosophy, A. F. Obolenskii, whom we regarded as cunning and unreliable (although that does not mean that he really was). We also disliked the professor of Mathematics and Physics, Iakov Ivanovich Kartsov, who suffered from dropsy which was in an advanced stage when I was promoted to the upper form. Moreover, this made him extremely irritable. Before this he was much respected and liked in the Lyceum. In such a moment of irritation he treated me quite cruelly. I had been sick and had spent several days in the hospital.* After recovering I returned to the Lyceum and my first class was in physics. I took my seat next to Sabir with whom I shared a desk. Kartsov had already begun his lecture. Not knowing what it was about, I turned with a question to Sabir who was given to clowning. He made me laugh and Kartsov heard me. Whereupon he called me to the blackboard and asked me to repeat his explanation. I told him that I could not do it because I had just returned from the hospital and did not know the class assignment. Kartsov began to question me on the previous lesson. This, too, I was unable to answer because I had also missed it. Disregarding my explanations, Kartsov announced that he was giving me the grade of zero for the month. He did so and to my great misfortune he never returned to the Lyceum

---

* During my entire stay at the Lyceum I was sick only twice; with measles when I was in the junior course, and the second time, the one of which I am now speaking, when I caught cold.

[which meant that the grade could never be changed]. He was bed-ridden and died about two months after this memorable incident.

For many months we had no physics class. Nevertheless, the cursed zero continued to be among my grades. The matter was serious because all monthly reports of grades in the upper form were taken into consideration in determining the graduate's rank or class. To receive the rating of ninth rank or titular counselor the student had to have a monthly average grade of 10 in the last three years. How many extra points did I have to gain in other subjects to make up the grade of 10 with a zero in physics for so many months! It seemed to me to be most unfair that I should be thus judged since there were no classes in physics and I had no chance to correct my poor grade. I went with my difficulty to our inspector, Obolenskii, and asked him to give me an opportunity to take a reëxamination in this subject to remove the undeserved zero which portended such serious consequences on my graduation from the Lyceum. Obolenskii, however, refused to grant my request. And all this trouble just because of a moment of laughter in class!

Kartsov was replaced in our Lyceum by Professor Shcheglov. Soon after that, our most kind Baron Egor Vasil'evich Vrangel' left us. After accepting the post of inspector of classes in the Imperial Institute of Jurisprudence, he was unable to leave Petersburg. He was replaced by Iakov Ivanovich Barshev, professor of Jurisprudence, who lectured in such a monotonous voice that he used to put us all to sleep. I must confess that I preferred horseback-riding, which I took from Gauer in the Court Riding School, to Barshev's lectures.

Since I was allowed more freedom in the upper classes than in the lower forms, I had an opportunity to make several new acquaintances in Tsarskoe Selo, particularly among the court hussars. Among them was Colonel Fedor Vasil'evich Il'in who had the reputation of being a great patron and friend of the Lyceum students. He used to come daily to our upper class during the recreation hours to have a game of chess with one of us. Occasionally I also used to meet the gifted poet Lermontov, who later became famous. He was at the time a cornet in the hussar regiment and used to lead quite a dissipated life. In this respect, however, he was outstripped by Lieutenant Alopeus, who was deported to the Caucausus for his inexcusable pranks! Lermontov followed him there very soon afterward.

Speaking of Lermontov, I also recall Pushkin who was in Tsarskoe Selo in the summer of either 1835 or 1836. As an old Lyceum graduate he used to visit us occasionally, and we always welcomed him with enthsuiasm.

I met Pushkin under the following circumstances. I now no longer spent my vacations with the Sul'menevs but instead with my brother Alexander, who lived with the Engel'gardts. One day my brother and I went with Sofia Grigor'evna to the French Theater which, thanks to the Engel'gardts, I patronized very often. As was our custom, we arrived early. The box next to us was still unoccupied. Suddenly the door opened and Pushkin entered with his beautiful wife and her sister, Miss Goncharov. The ladies took the front seats and Alexander Sergeevich sat in the back and, since I occupied the same position in my box, we found ourselves side by side in the closest proximity, separated only by a rail. Glancing at my uniform Pushkin said to me: "So you are a Lyceum student!" After I answered in the affirmative, our talk began. During this unforgettable conversation I had occasion to call him by his name. "You know me then," said Pushkin. "I would be no Lyceum student if I did not know you," I replied. Alexander Sergeevich smiled and became even more affable toward me. Thus I sat an entire evening next to the gifted poet. He seemed to be in very good spirits and laughed heartily over the farces of Vernet and Paul Minet. He applauded the actress, Mme. Bras, a great deal. But I just cannot recall the name of the play. During one of the intermissions Pushkin invited me to the refreshment bar and bought me an ice cream.

At that time the Engel'gardts occupied a magnificent apartment in the house of Shcherbakov (later Admiral Melikhov) on Mokhovaia Street. Everything was sumptuous in their home—the furnishings, the servants, the table, the carriages, and the rest. They had a permanent box at one of the theaters, and I always used it with great pleasure. However, they received few people at home. Paul Vasil'evich seldom went out. He remained all day in his study in his dressing gown. In the evening he went to the theater, but not to his wife's box. He had a reserved seat of his own. Sofia Grigor'evna too had no other pleasures than the theater. The relations between husband and wife did not seem to be bad, but very strange. Apparently they gave each other complete freedom to do as each pleased. My brother Alexander established a great friendship with Sofia Grigor'evna and spent all his leisure in her company. He became quite a stay-at-home, and it was with great difficulty that I could persuade him to visit the kind Sul'-menevs where, as usual, our relatives gathered. In order not to accept without payment the hospitality of the Engel'gardts, to whom he had become greatly attached, my brother began to teach their children (they had four small sons) every day. These little boys had a young German, Shul'ts [Schultz], whom we called simply "Jean," who could not pretend to be a tutor because of his education. He was a sort of supervisor or male nurse who took care of the children. He was honest, had a fine character, and was an intelligent young man. I became quite friendly with him and could always depend on him when I needed a favor.

My brother began his service in the Chancery of the State Council, but when the Ministry of State Properties was formed under the administration of General Kiselev, he was transferred there and was appointed section chief of the Cadastral Division of the Third Department. The director of the latter was at first General Dellingsgauzen and later Privy Councilor Bradke. My brother, who was intelligent and able, took his administrative duties very seriously and was always praised by his superiors.

One of my Lyceum school friends, De Brin'i, also visited the Engel'gardts. He was Sofia Grigor'evna's nephew. When I was already in the upper form, another of Sofia Grigor'evna's nephews entered the Lyceum. His name was Count Gudovich. His mother came to live in Tsarskoe Selo, and I visited her quite often. She also had a daughter, Countess Nina Gudovich, a charming young lady of whom I lost track.

Thus I knew many homes in Tsarskoe Selo which I could visit on Sundays. However, during the stay of the court in Tsarskoe Selo I always visited the Litkes. My kind and charming aunt Iulia Vasil'evna lived in Tsarskoe Selo from May to the end of Autumn. Here she had much more comfortable quarters than in Peterhof or in the other places where she had to follow the tsar's family. I always recall with a feeling of gratitude her genuine concern for me and the attention she showed me. My aunt Baroness Elisaveta Petrovna Rozen usually stayed with her. She left the Institute after her daughter Matilda and also my sisters Emilia and Valeria graduated. During the first year of her marriage Iulia Vasil'evna occupied an apartment in the Alexander Palace, but later she lived permanently in one of the pretty little houses of the Chinese Village. Sometimes I used to meet the grand duchesses in her home, particularly the late Alexandra Nikolaevna who, it could be said, simply adored her. One very warm day we all sat in the summer house. Uncle Fedor Petrovich became so warm that he took his coat off and put on a white caftan. Suddenly we heard near the summer house a noise and calls: Julie! Julie! It was the grand duchesses who were coming straight into the summer house. My uncle was quite embarrassed; he tried to jump into the bushes, but got entangled in his long coat and almost fell. The grand duchesses burst into loud laughter at this scene. My uncle ran away. We, however, remained for a long time and enjoyed ourselves immensely.

At the end of 1836, or perhaps early in 1837, I received the news of the engagements of both my sisters; Emilia to Ivan Petrovich Speranskii and Valeria to Nikifor Ivanovich Anferov.

Before leaving for Kamenets with my two elder sisters my mother first stopped in Dubno and Radzivilov to conclude her affairs and also to pick up my younger sisters Iulia and Anneta, who had remained with their sister Sofia Sitnikov. Speranskii, who served in the Department of Foreign Trade, happened to be in Radzivilov at the time and proposed to Emilia. Anferov was a public prosecutor in Kamenets where he met my sister Valeria. After this double wedding which, of course, I did not attend, the Anferovs remained in Kamenets and the Speranskiis left for Berdichev where Ivan Petrovich had a special and long assignment. Soon after that N. I. Anferov was transferred to Zhitomir as public prosecutor where he remained for a long time.

I spent my vacation in 1836 with my brother Alexander, in other words with the Engel'gardts who were living at that time on Aptekarskii Island near the Karpovka. The Sul'menevs also lived in a dacha not far from them, which rejoiced me greatly because I could spend most of the time with them.

On my return to the Lyceum after my vacation I continued to visit the Litke family as long as they remained in Tsarskoe Selo, that is, until the return of the imperial court to Petersburg. During all this time I seem to have been assigned more frequently than before to outside activities such as guard duty during the presence of the tsar's family in church or to attendance at the balls given in the palace. The effect of these amusements on my studies was not too favorable. Nevertheless, I did not lag far behind and, after passing the examinations successfully, I was promoted, seventh or eighth in the class, to the fourth or graduating class.

I have already said that ours was the ninth graduating class since the foundation of the Lyceum. We were for some reason unable to get along with the eighth class which preceded us and we were not on speaking terms for a long time, but just before graduation we called a truce and fraternized. Incidentally, among the members of the eighth graduating class were Baron Ikskul von Guldeband, at present our ambassador in Rome, and the unfortunate Count Lev Sivers, who for almost thirty years remained in the post of secretary to our mission in The Hague. He has now been appointed consul-general in Amsterdam.

Our promotional examinations were just before Christmas of 1836. On finishing them we were given our swords and were allowed to go home. I must confess that this new ornament was a source of considerable delight to me. It placed me, as I then imagined, in the ranks of the people of some consequence. At first, when among people I used every pretext to take off and put on my sword, hoping, of course, to attract attention to myself. It goes without saying that this only provoked derisive smiles on the part of those present. But who in his youth has not experienced this feeling of petty vanity which in mature years seems so ridiculous?

Shortly after my return from Christmas vacation, at the end of January 1837, we received the sad news of the tragic death of Pushkin of whom we were so justly proud. We were terribly upset by this misfortune and were indignant at the foreigner D'Anthes, the assassin of our immortal poet. A manuscript was being passed among us from hand to hand then of a suppressed poem by Lermontov, "The Death of the Poet," [8] and for some time one could hear the furious reciting of these inimitable verses in our halls.

If I am not mistaken, that same winter we were stunned by a dreadful but magnificent sight. Returning one evening after prayers to our dormitories, which were on the top floor, we saw through the windows a tremendous glow coming from the direction of Petersburg. The blaze increased and became unusually bright. No one seemed to be able to explain this phenomenon which disturbed us for a long time. On the following day we learned that the blaze came from the Winter Palace which was on fire. The fire was so great that the glow could be seen very clearly at a distance of twenty-five versts!

Another incident dates to this time, less important than the death of Pushkin or the fire in the Winter Palace, but one which embarrassed me to such an extent that I cannot think of it to this day without shame. Our kind General Gol'tgner gave a ball on the occasion of some celebration in his family. He invited several Lyceum students including myself. After the dance which I enjoyed immensely, we were invited to supper. I sat at a small table occupied by several hussars and cuirassiers, also by Gol'tgner's son, at that time a captain of the Paul Guard Regiment. We were all very gay. Suddenly my table companions decided to get me drunk, and I do not know why this silly prank proved so successful, but I passed out and rolled under the table. All attempts to put me on my feet again were in vain. I was carried home and put to bed. The following morning our director came to visit me. He reprimanded me mildly and advised me to be more careful in the future. I did profit by this lesson and nothing like it has happened to me since.

During Lent of 1837 our pastor Avenarius prepared the Lutherans of our class, Gernet, Gol'tgner, Baron Ikskul, and myself, for confirmation or first Holy Communion. I must give him credit for fulfilling this duty so conscientiously and with such Christian devotion that it was with awe and full awareness of the significance of the sacrament that I approached the altar. This Lutheran custom of preparing young people for confirmation when they are old enough to grasp fully the true importance of the act has much to recommend it. However, I am convinced that the Russian Orthodox Church is wiser in admitting children also to the bliss of the sacrament.

My vacation trips to Petersburg during 1837 were memorable to me because of my frequent attendance at the theaters. At that time our Russian stage was at a height which it has never since attained. The Karatygins, Branskii, Dior, Sosnitskii, Mme. Asenkov, and particularly Martynov were at the apogee of their fame. The Samoilovs were just making their debut then. The French theater was also at its prime. There was no Italian opera yet. However, at that time the Russian and German opera companies were not bad. Their reputation dropped considerably with the appearance of the Italian singers. Practically all operas by Rossini, Bellini, Donizetti, Auber, Meyerbeer, and others were then novelties. I was present at one of the first performances of *Fenella* and *Robert the Devil*. I witnessed and participated in the excitement produced by these and similar plays in the audience. It was at that time too that Glinka's opera, *A Life for the Tsar* [*Zhizn' za Tsaria*], made its sensational appearance on the stage. Everything is different now. Seldom does anything fresh appear in the repertoire, and if a new composition is offered, one has to listen to it many times to understand and retain it. And how excellent our dramatic theater was! What could compare with *The Inspector-General* [*Revizor*] which I saw almost as soon as it made its appearance on the stage! The same can be said about our Russian literature in general. Such writers as Pushkin, Zhukovskii, Lermontov, Kozlov, and Baratynskii were producing their best works. And even Krylov was still writing. One could say that hardly a day passed that some remarkable work was not published! The monthly magazine, *The Reader's Library* [*Biblioteka dlia Chteniia*], in which Senkovskii under the pseudonym of Baron Brambeus published his remarkable works, was read with great interest. Only the newspapers and the political literature were very weak. Moreover, it could not be otherwise under the existing regime. At present all our activity is directed toward the development of industry and our material interests. People find inspiration in discovering some new machine. Material benefits are pursued. How can poetry and the fine arts flourish! Imagination is dead. But is mankind happier? It is very questionable.

I spent the summer vacation of 1837 again with the Engel'gardts in the dacha which they bought from Seliavin and where I had met them two years before during my mother's stay in Petersburg. I do not recall where the Sul'menevs lived then. Apparently somewhere far away, because I visited them seldom at that time. Life in the Engel'gardt's home was luxurious, but rather sad because I met there people who were strange to me and with whom I had little in common. I missed my relatives. Therefore I returned to the Lyceum without regret. From there I used to visit the Litke family to whom I felt a great kinship. They lived at that time in the Chinese Village in the center of my beloved Tsarskoe Selo garden. I think that it was this year that their elder son Konstantin was born. He was a charming child whom I grew to love.

We were dismissed on Saturday after classes and had to return to the Lyceum by 10 o'clock in the evening. On Sundays we were dismissed after mass which all had to attend. I had to go to the Lutheran Church except on the days when I was on duty with the tsar's family.

One Saturday evening on returning from the Lyceum I found the kind Iulia Vasil'evna with tears in her eyes looking very sad. This surprised me the more because she was of a calm nature and knew how to control herself. In spite of her usual gentleness toward me, our conversation seemed to lag, whereupon I took the liberty of asking her the cause of her sadness. She then burst into tears and told me that she was greatly worried about my uncle. For some time, she said, he had grown morose and irritable and seemed displeased with everything. Despite all her efforts, she could not make him tell her the cause of such a change in his mood. Although he never complained about his health, she was convinced that he suffered both physically and morally. In spite of his great love for his wife, whom he respected as well, Fedor Petrovich because of his peculiar nature almost never shared with her his innermost thoughts on questions that did not bear on their family life and interests, and if any unpleasantness happened to him, he never discussed it with anyone.

Iulia Vasil'evna suspected, however, that her husband was having serious difficulties in tutoring the willful grand duke, a task which Fedor Petrovich had undertaken so unwillingly.

Her suspicions were confirmed. Several days after my conversation with my aunt, the Grand Duchess Alexandra Nikolaevna, who came to see her, told her that she overheard a conversation between the tsar and the empress during their carriage ride. The empress complained that Fedor Petrovich treated the grand duke too cruelly and begged the tsar to replace him with another tutor. The tsar would not agree; he praised the high qualities of Litke who had his entire respect and confidence and he tried to prove to the empress that it was necessary to have a firm tutor like Litke with a willful boy such as the Grand Duke Konstantin.

My aunt naturally passed this secret on to Fedor Petrovich, but he received it indifferently and in no way changed his treatment of the grand duke despite the frequent interferences on the part of the empress. I shall give a few instances which were told to me by Iulia Vasil'evna.

Once the empress sent word to Litke to excuse the grand duke whom she wished to take for a carriage ride. Fedor Petrovich replied that His Highness had a lesson. The courier returned to say that the empress ordered him to postpone the lesson for an hour. Litke replied that another teacher was coming to give the grand duke a lesson in an hour, and that, consequently, he could not excuse the grand duke. He won his point.

On another occasion the grand duke had the idea of taking piano lessons and began to ask Fedor Petrovich to get him a teacher. My uncle replied that he should first think seriously about the boring difficulties of first music lessons and that after he had taken a firm decision to overcome them and to keep up his lessons, only then would my uncle engage a teacher for him. Moreover, Litke warned the grand duke that once the music lessons were begun he would not permit him to stop them, no matter how dull they might seem, because a work well considered and started should never be abandoned; however, to jump from one thing to another senselessly was not only useless but even detrimental. The impatient little boy bothered Litke every day, begging him to get a music teacher as soon as possible. Seeing that Litke kept postponing action, the boy decided to complain to his mother. On the first occasion the empress told Fedor Petrovich that she wanted the grand duke's wish granted, but even now Litke refused to yield and succeeded in

gaining her consent to wait a little longer to give the grand duke
an opportunity to weigh his decision which he would have to abide
by once the lessons began.

Finally after this long probation a teacher was engaged and
hours were arranged for the music lessons. At first all went well,
but soon the little boy grew bored and pleaded that the lessons be
stopped. Litke, of course, did not consent. Scenes, tears, and finally
new interference by the empress followed. Nevertheless, Fedor
Petrovich remained adamant despite the obvious displeasure of the
empress which was hard for him to bear. The emperor trusted
Litke implicitly and gave him almost unlimited freedom in the
education of the grand duke, but this time he sided with the empress. He summoned Fedor Petrovich and said that he absolutely
failed to understand why the grand duke was forced to study music
when he had no particular inclination for it. Litke submitted his
arguments to His Majesty and added that in all conscience he could
not yield in this matter. "Perhaps I was wrong," he added, "in the
understanding of my duties as a tutor for which I have never had
any particular inclination; in this case, I beg Your Majesty to dis-
miss me from this post."

The arrogant but just Emperor Nicholas was by no means
angered by this explanation. On the contrary, he embraced Litke,
thanked him for his frankness and for his faithful service, and
asked him to continue just as conscientiously as before. The music
lessons continued and the grand duke learned to play the piano
quite well.

Needless to say, such difficulties recurred repeatedly. They naturally affected the disposition of Fedor Petrovich who was seldom
in a good mood. During the time that he had charge of the grand
duke's education one could seldom see him cheerful. Later, when
the education of the grand duke was completed, he heaved a sigh
of relief and became the same amiable, warm, and cheerful man
he used to be.

Because of my relations with the Litke family, I had frequent
opportunity to meet the Grand Duke Konstantin Nikolaevich, and
I could to some extent follow the rapid development of this unusually able boy. As early as that time a favorable opinion prevailed
among the public about his remarkable gifts and an unflattering

one about his character. The opposite opinion existed about his elder brother, the presently reigning tsar, who was not particularly gifted, but who had a kind and soft heart. No one suspected that his reign would be one of the most brilliant and beneficial ones for Russia. The mind of the heir began to develop in his adult years, and the fine admonitions of his wise and noble guide, the unforgettable Zhukovskii,[9] nurtured his good qualities and raised them to the height to which not only Russia but the entire enlightened world has paid just tribute.

In 1837 the Grand Duke Mikhail Pavlovich again went abroad, and our temporary superior was the young prince Peter Oldenburg. He was good to us and did not bother us as did Sukhozanet. The grand duke was absent only a few months. On his return to Russia he began to visit us as before. I recall that once he told us that while he was in Rome attending some conference, his top coat was stolen and that he was chilled returning home. And now, he said, "some crook no doubt is charging a fee for showing the coat of a Russian grand duke as a curiosity."

Less than a year remained until my graduation.* In our correspondence my mother and I made plans for my future. I began to think about entering military service, but my mother suggested that this question be decided on her arrival in Petersburg, where she planned to move by the time that I graduated from the Lyceum. In the meantime my grandfather Fedor Ivanovich Engel offered to use his influence with the famous Konstantin Konstantinovich Radofinikin, with whom he was on very friendly terms, so that I could enter the Ministry of Foreign Affairs. Radofinikin continued at that time to have charge of the Asiatic Department although he had long been a member of the State Council. Diplomatic service, particularly in the east, attracted me, but so did a military career. I was undecided, and resolved to postpone a final decision until the arrival of my mother. Soon after that, at the end of 1837 or early in 1838, Fedor Ivanovich Engel died at an old age, and a month before my graduation from the Lyceum Radofini-

---

* That year, at my mother's wish, a portrait was made of me in the Lyceum uniform of the upper form. After my father's death it was kept in the home of my sister Valeria, who later gave it to me.

kin died also. Thus I was prevented from taking advantage of recommendations from these statesmen.

My uncle Engel was a man of remarkable mind and education. I have spoken about his career earlier in these notes. In the later years of his life his sphere of activity was limited to his duties in the State Council where he presided over the department dealing with the affairs of the Polish kingdom. He was also present, of course, in the general session of the council, and he frequently regretted that there were so few enlightened men among his colleagues. Once on leaving a sitting at which the numerous debates among the members on some important state question were not resolved, he turned to N. N. Novosil'tsev,[10] one of the most remarkable dignataries of his time, and said, "*C'est prodigieux ce qu'ils ignorent.*" Unfortunately the same thing could be said now about most members of this chief legislative institution of Russia. Let us hope that the great and beneficial reforms of the present reign will also include this institution.

After Christmas and New Year, which I again spent with my brother Alexander in the home of the Engel'gardts, we began intensive preparations for the final examination at the end of April. However, in the midst of these studies, no more than a month before the examination, I received the terrible news of the death of my dear and adored mother. It was transmitted to me through my brother Alexander who came to Tsarskoe Selo to share our mutual grief.

My mother was already on her way to Petersburg where she planned to settle with us, as I have already said. She stopped on the way in Zhitomir with my sister Valeria whose husband was then provincial public prosecutor. My sisters Iulia and little Anneta were with my mother. Iulia was beginning to go about in society. Although not feeling very well, my mother accompanied her to a ball. In the vestibule the attendant could not find mother's fur coat for a long time and she became chilled in her light gown. Immediately on her return home she felt feverish. Her illness took a turn for the worse, and, on February 21, 1838, my treasured mother was no longer. She died at the age of forty-four years of her virtuous life. Truly the ways of God Almighty are impenetrable! Both of our parents, whose sole happiness was to see their children

succeed were struck by death at the moment when they could enjoy this happiness and be reunited with their children after a long
separation. My mother's body was committed to the earth in
Zhitomir. My brother Alexander went there immediately to take
care of our younger sisters. He took Iulia to Kamenets to our uncle
Fedor Karlovich, whose daughter Valeria was almost of the same
age, and he brought Anneta to Petersburg where she stayed at
first with the Engel'gardts. Later Iulia Vasil'evna Litke took her,
and, on the death of the latter, Anneta entered the Catherine Institute.

I shall not attempt to describe the terrible sorrow and grief that
I felt at the loss of my mother whose arrival I had awaited so impatiently. It seemed to me that nothing mattered now that my
hopes of being reunited with her in our family circle in Petersburg
had been shattered. My sad mood was, of course, not conducive to
the preparation for the final examinations.

Shortly after this an unpleasant incident occurred whose consequences might have affected the entire future of our class.

Our inspector, Andrei Fillipovich Obolenskii, with whom we
never got along, decided to reprimand my schoolmate De Brin'i
for wearing his hair too long and ordered him to have it cut. We
were practically graduates and foolishly considered ourselves above
such orders. De Brin'i, who felt insulted, told the inspector in a
rather rude manner that he did not intend to follow his order.
Angered by such a reply, Obolenskii called several maintenance
men and with their aid put De Brin'i in the school detention room
—a punishment almost never used even in the lower classes. Our
entire class was enraged. We were at the time having tea. Obolenskii was showered not only with insults, but with rolls and
spoons, and, in short, with anything we could put our hands on.
Leonid Voeikov even threw a decanter at him. The inspector fled
and we went to free De Brin'i. Fortunately, in his solitude he had
had time to think the whole matter over and to realize the danger
of our act. He refused to leave the detention room before the return of the director, and he begged us to calm down. On learning
about the disturbance, our poor director General Gol'tgner arrived
greatly alarmed to reprimand us. We faced him in silence because
no one felt any animosity toward this venerable old man. With

tears in his eyes the general announced to us that our case was very serious and must be brought to the immediate attention of the tsar himself. Obolenskii, in the meantime, had departed for Petersburg to make a personal report to Grand Duke Mikhail Pavlovich.

On the following day the storm broke. The senior adjutant of the grand duke, Colonel Iakov Ivanovich Rostovtsev, arrived at the Lyceum and ordered us to line up.* After giving us the worst scolding possible, he read solemnly the supreme order of the tsar according to which Voeikov and two of our other schoolmates, whose names I do not recall, were demoted to the rank of private, and the rest of us to that of military clerk to be sent to various remote provinces. De Brin'i, however, was forgiven for his honorable act and was the only one allowed to take the final examination. Then, turning to our director, Rostovtsev said: "And you, general, the tsar orders you arrested for your unpardonable weakness. Your sword, please!" At these words and on seeing our venerable director with his head sadly lowered, we all rushed to Rostovtsev and began to implore him not to pardon us, but to deliver our kind director from any punishment. The scene became quite emotional, particularly when De Brin'i announced that he did not care to be an exception and wished to share the fate of his schoolmates. Deeply touched by this scene and our pleas, Ia. I. Rostovtsev said that he would report everything he had witnessed to the grand duke and would act as our solicitor.

God sent us another warm solicitor in the person of Countess Bobrinskii, who resided at that time in Tsarskoe Selo in order to be near her son, a student at the Lyceum. On learning of our misfortune and after obtaining the details of what had happened, she hurried to Petersburg without losing a moment's time to see Grand Duke Mikhail Pavlovich who was a good friend of hers. She returned the following day to Tsarskoe Selo with the good news that the grand duke hoped to save us. But it was only a hope, and we spent the following few days in nervous anticipation.

---

* Ia. I. Rostovtsev was later chief of staff of the military training schools and became influential under the presently reigning tsar. He was adjutant general and was elevated to the rank of count in connection with the abolition of serfdom with which he was entrusted.

Finally Rostovtsev arrived with the glad news that we were all pardoned by His Majesty. Our kind director was elated and, of course, we rejoiced no less.

Then the examinations came. They passed satisfactorily but to our shame be it said not without cheating in some subjects. One of us succeeded in marking many of the question tickets so that they could be identified. I do not know whether the school administrators noticed it, but, since many outside persons were present at the examinations, they apparently did not think it appropriate to control us too much. During one of those examinations, when I was called, I noticed to my surprise my kind uncle Ivan Savvich, who listened to me with attention. The Sul'menevs were spending that summer in Tsarskoe Selo.

Finally the day of our graduation arrived. I do not remember exactly what day in May of 1838 it was. My brother Alexander called for me and brought my civilian clothes which had been ordered from the tailor Klemens a long time in advance. I must say that we felt pretty awkward in those suits which were made according to the prevailing style so tight that we could barely get into them. And what colors! Some were green, some blue, some brown. The vests and ties were particularly brilliant which was then considered fashionable. I was the only one in black because of mourning. Our greatest difficulty was the haircut. Enormous waved tufts were then the fashion for men which meant running to the hairdresser almost daily. We had two or three such artists to take care of our hair.

Thus dressed we went to bid farewell to our administration and to our friends in Tsarskoe Selo. I recall how my kind aunt Natalia Petrovna laughed on seeing me all dressed up, and, to be sure, one needed considerable time to get accustomed to this hairdress and wear it casually. I made the same impression on the kind Litkes who said that I could be mistaken for an actor. However, this comic element in the situation did not in any way prevent my dear relatives from taking the sincerest interest in me and from wishing me all the happiness in my future career, upon which I was entering too early, because I was barely eighteen years old.

That same day I learned to my chagrin that I had graduated only tenth in my class, that is in the rank of collegiate secretary and not of titular counselor as I expected. To console me General Gol'tgner told me that I was first in the tenth rank, but in this case I did not share the opinion of Julius Caesar that it is better to be first in the village than second in the town! I felt that it was not fair that some of my classmates who were not as good in many subjects as I were placed higher. This, as it was revealed, occurred because of the zero grade given to me in a fit of anger by the professor of Physics, Kartsov, a grade which remained for a long time in the monthly reports, thus reducing the average necessary for the rank of titular counselor. Thus my inappropriate laughter in the classroom cost me four years in the service!

Here is, as far as I can remember, the order in which we graduated from the Lyceum.

In the rank of titular counselor:

1. Veselovskii
   received gold medal
2. Kaidanov
   received gold medal
3. Arsen'ev
   received silver medal
4. Galitskii
   received silver medal
5. Akhsharumov
6. De Brin'i
7. Nekliudov
8. Baron Bernard Ikskul
9. Iakhontov

In the rank of collegiate secretary:

10. Giers
11. Sabir

In the rank of provincial secretary:

12. Obukhov
13. Gernet
14. Gaberzang
15. Stessel'
16. Zakharzhevskii
17. Trofimov
18. Teplov

In the rank of collegiate registrar:

19. Ginter
20. Voeikov

Ours was, without a doubt, the most unfortunate graduation class since the founding of the Lyceum. Six of our classmates failed to graduate with us. Bogaevskii and Kreiton had to leave the Lyceum because of illness, and soon died. Another Bogaevskii transferred to some military school. The younger Voeikov (Alexei), Marchenko, and Petrashevskii remained with the previous class to repeat.* And, if we follow the further fate of my class-mates, we shall see that only a minority remained alive and among them very few left a mark in the world.

My brother and I went to Petersburg by the railroad which was opened not long before my graduation. This first railway built in Russia aroused universal curiosity and amazement. My brother and I settled down with the Engel'gardts who lived at that time in their own dacha on the Karpovka, near the garden of Count Laval.

* In place of them we acquired Gaberzang who remained in the third class of the preceding course, and thus became our classmate.

IV

SERVICE IN THE ASIATIC DEPARTMENT
(1838–1841)

I PASSED QUIETLY and even sadly the first months of my entrance into society, the months which usually constitute the happiest period in the life of a young man. My deep mourning for my mother prevented me from seeking worldly amusements. Moreover, where could one find them? The circle of acquaintances of the Engel'gardt family was limited and consisted of people who interested me little, and practically none of my relatives, whom I loved and who were close to me, were in Petersburg. Occasionally I went to Tsarskoe Selo to visit the Sul'menevs and Litkes. I used to spend my days in reading French novels or in taking boat trips on the Karpovka or the Little Neva near the garden of Laval where I often walked. However, I enjoyed, of course, some pleasant moments living together with my brothers (Fedor was already in the School of Jurisprudence and spent his vacations with us) and my little sister Anneta who remained with the Engel'gardts for some time. I became close friends with Jean, whom I have already mentioned in my memoirs, although he was a man of little education. However, he possessed a beautiful soul and a most honorable character. He also had a remarkable practical intelligence. I talked a great deal with him about my future. He tried in every way to discourage me from choosing a military career about which I still dreamed. He tried to persuade me to enter the Ministry of Foreign Affairs and to try to succeed in the East which offered greater activity and, consequently, opportunities for a diplomatic career for a young diplomat. It goes without saying that my brother Alexander, who took the keenest interest in my future, also gave me some good advice.

Adrian Vasil'evich Bykov, the husband of Nadin'ka Sul'menev, had been serving for some time in the Ministry of Foreign Affairs. He offered to recommend me for admittance to this ministry. He was chief assistant to Privy Counselor Bek, who had charge of the Secret Expedition where codes were prepared and employed. Adrian Vasil'evich proposed that I begin service in this Expedition. To this end he began to teach me the codes, having first secured my promise that I would tell no one the secrets which would be confided to me. He began his instruction by deciphering with me a dispatch or, to be more exact, a denunciation by some agent in Paris against one of the most respectable men of high standing in Russia. While deciphering the denunciation I became so indignant and felt such loathing that I declined most definitely to serve in the Secret Expedition. I assured Bykov once more that I would keep my promise not to tell anyone of the content of the denunciation revealed to me and I told him that I wished to remain true to my first intention of entering the Asiatic Department.

After the recent death of the wise and shrewd Radofinikin this department was administered at that time by the vice-director Lev Grigor'evich Seniavin, a young official of twenty-seven who was particularly favored by Count Nesselrode. Rumor had it that the count wanted to see Seniavin marry one of his daughters. By his birth and connections Lev Grigor'evich belonged to the aristocracy, but he disliked worldly pleasures. Despite his title of *Kammerherr*, he practically never appeared at court, and his only pleasure seemed to be his service in the department. He was an official of the vain and arrogant type in a high office. He was not devoid of some ability, but his horizon was extremely limited. He did not delve deeply into political questions, and he looked upon them from the formalistic point of view. Nevertheless, his department was in model order; his officials were held in strict discipline, and all paper work was punctual. Before his promotion Lev Grigor'evich was courteous and attentive toward his subordinates, but later he became quite unpleasant in his treatment of them and everybody was terribly afraid of him. However, he did have some good qualities. At any rate, he deserves credit for his noble character and the dignity with which he held his position.*

* He was a brother of Ivan Grigor'evich Seniavin, deputy minister of the interior. Both of them were unusually tall. Lev Grigor'evich was a bachelor; his brother, married to a member of the Dev'er family, had a large family.

In June I was introduced to Seniavin by Bykov. Since every Lyceum graduate had the right to select the service he wished, there could be no objection to my admission to the Asiatic Department even if no vacancy existed there at the time. And until such a vacancy occurred, each Lyceum student would receive from the state treasury a yearly salary of 800 rubles in bank notes. For this reason Lev Grigor'evich agreed to employ me and, in general, treated me courteously. I did not know, however, what his motives were in suggesting that I should not report to the department until early in October. I was listed officially as having entered the service on October 13, 1838! In my introductory interview with Seniavin I expressed a desire to remain in the department for some time to familiarize myself with the administrative aspect of diplomatic service and that later I should like to be sent abroad to the East. This wish was fully approved by Lev Grigor'evich.

On arriving in the city with the Engel'gardts at the end of September (Mokhovaia Street, the house of Shcherbakov), I began my preparations for my work in the department. I ordered the uniform of the Ministry of Foreign Affairs. This uniform was respected then as much as that of a *Kammerjunker*. I reported to the department on the appointed day. After taking the usual oath in the chapel of the ministry, Lev Grigor'evich showed me the place I was assigned. It was in the first room as one enters the department, next to the office of the director.

The Asiatic Department consisted then of three divisions. The first was the Turkish-Greek Division and was headed by Vasilii Sergeevich Lapin. The second division dealt with the affairs of

---

Although they had considerable means, Ivan Grigor'evich was soon practically bankrupt because of the extravagance of his wife. His brother helped him, and in order to do so himself led a modest life. But this was not enough to extricate Ivan from his financial difficulties. Therefore Lev Grigor'evich decided to sell his home on the English Embankment and his magnificent estates in the province of Voronezh. This honorable and generous act failed, however, to save Ivan Grigor'evich. His affairs reached such a state that in a moment of despair he cut his throat with a razor and threw himself out of a third floor window. This tragic accident which occurred early in the 'forties made a strong impression on Petersburg society. Lev Grigor'evich took the entire family of his unfortunate brother under his care.

China, Central Asia, and Persia, with Nicholas Ivanovich Liubimov at the head. And the third, or Inspector's Division, dealt properly with the clerical aspect of the department, such as the keeping of lists of employees, recommendations for awards, making out of payrolls, and the like. Paul Ivanovich Skatov was in charge of this division. Each division had two desk heads. I recall among them Ivan Dmitrievich Khanchinskii, Timkovskii, Khrushchev, and Zlobin. Only these persons and also Baron Jomini, who was attached to the department to handle minor French correspondence, edited papers. All other officials, including the assistants to the desk heads, merely copied papers—in other words, did the work of ordinary copyists in other ministries. I found myself among them. This chagrined me the more because my handwriting was poor and, as a result, I was seldom given any political dispatches to copy, but they always shoved the routine office papers my way. In spite of this I worked assiduously without complaining in the hope that in return I would receive more interesting work. In the meantime I tried to improve my knowledge by reading some of the political correspondence. However, since Lev Grigor'evich was strict about keeping this correspondence secret even from us, I succeeded in reading it only stealthily and that thanks only to my friendship with our old record keepers Shilovtsov, Naumov, and the expeditioner Veikardt. Even these venerable persons, contemporaries of Radofinikin, lived in constant fear of incurring Lev Grigor'evich's wrath who but recently had been their subordinate.

I am greatly indebted also to my friendship with Iakov Danilovich Ginkulov, a native of Bessarabia, who as dragoman of the Moldo-Wallachian language in the department had charge of the entire statistical and political archive which he kept in exemplary order. At the same time he taught Moldo-Wallachian in the Oriental Department of the University of St. Petersburg. Iakov Danilovich was an intelligent and honest man. Seeing my curiosity and keen interest in the political affairs of the East, as well as my desire to master them thoroughly, he gladly initiated me into these matters and gave me good advice. He thus became my first instructor in the field of my choice. He treated everything seriously and subjected to critical analysis every circumstance no matter how trivial it appeared. A great admirer of Radofinikin, under whom he had served for a considerable time, he always spoke with particular respect about this outstanding statesman. And indeed when reading Radofinikin's correspondence in the archives, particularly the letters written in his own hand, I never ceased to wonder at his insight and the correctness of his views. In one of these letters to Griboedov,[1] the former diplomatic representative in Persia, he condemned his actions and virtually predicted the catastrophe which befell him. I recall particularly the following sentence in this prophetic letter, mailed not long before the death of the famous author of *Gorie ot Uma* (*The Misfortune of Being Clever*), "and a good share of the guilt will be on your conscience." Radofinikin had an excellent command of the Russian language and, except for the dispatches signed by Count Nesselrode, the entire correspondence, even that which was purely political in content, was carried on in Russian.

Among my co-workers who sat in the same room where I was, I recall Prince David Golitsyn, Baron Istrov, Debu, Konstantin Ivanovich Lents, Pizani, Bade, Reingold't, Vranov, Ol'khin, and Pavlov, who joined the department long after me and whom I later met in Persia. Because of their excellent handwriting many of them enjoyed the particular favor of Seniavin, who constantly gave them awards. However, this did not open the road to a future career to any one of them.

The more important political papers, even those on the affairs of the East, were always written by the counselors of the ministry, Baron Brunov, Baron Osten-Saken, Ladenskii, and Ivan Sergeevich Mal'tsev. These gentlemen kept going in and out of Seniavin's office where they received papers, and they in turn brought him dispatches written by them and already checked by Count Nesselrode. These were copied and filed in our department. After the appointment of Baron Brunov as ambassador in Stuttgart, our dispatches were handled for the most part by Ladenskii and later by Rikman and Fedor Furman with whom I had no time to become acquainted despite our kinship. I also recall frequent visits to our department by Privy Counselor Anton Fanton, at one time senior dragoman in Constantinople, and the chief participant in the negotiations on the Treaty of Adrianople. He was then attached to Vice-Chancelor Count Nesselrode, who liked and respected him and always consulted with him on matters of the East.

With the help of my friend Ia. D. Ginkulov, and some others, I frequently had the opportunity of reading our political correspondence and I particularly admired the despatches by Brunov. His remarkably clear, brief, and expressive style excited me! Only by this means was I able to gain any benefit from my service in the Asiatic Department. My official work there, however, was not only trivial, but dulled intellectual ability. It is understandable therefore why I was so anxious to go abroad, even to the most remote parts of the East.

In the meantime, in January of 1839 I received a promotion to junior assistant to the desk head. This added somewhat to my financial resources, but in no way satisfied my pride because my duties remained the same. Moreover, I continued to dream about the means of obtaining some appointment abroad.

At the end of the same year a new mission to Pekin was assembled. According to the treaty existing with China at that time we had the right to maintain there only a religious mission. In addition, it had to be changed every ten years. As is known, this mission was allowed by the Chinese government to accommodate the considerable Russian colony in Pekin, composed of Cossacks taken prisoner some time ago on the Siberian border at our fortress of Albazin.[2] The mission consisted of an archimandrite, two priests, an equal number of students from the Theological Seminary, and a doctor. Taking advantage of this replacement every ten years, our government always sent with the new mission a diplomatic agent, officially designated as a police commissioner. The Chinese government did not allow this police commissioner to remain in Pekin after the ten year period when the mission had to be replaced. This practice was continued until 1860 when General Ignat'ev was sent to China. He succeeded in concluding a treaty which gave us the right to maintain a permanent diplomatic mission in Pekin.

In 1839, the year about which I am speaking, our chief of the division, Nicholas Ivanovich Liubimov, was appointed police commissioner. For several months we were busy with the preparations

for this mission. They aroused in me a strong desire to join Liubimov and, without thinking too long about it, I went to Lev
Grigor'evich with my request. I did so just at the moment when
he was leaving the department and was putting on his top coat.
"Why? Do you want to become a monk?" he asked with derision.
I frowned and replied that I did not think that I had to be a monk
to be attached to the police commissioner and to accompany the
mission. Apparently Seniavin did not like my reply. Without saying another word, he jumped into his carriage.

I related this conversation to I. D. Ginkulov, who did not approve of my escapade. He mildly upbraided me for not consulting
him first. Because of my inexperience (I was not yet nineteen) I
did not know how careful one had to be in approaching so willful
and arrogant a superior as Seniavin. I should not have stopped him
in the doorway or in the cloak room, but should have come to his
office with my head bowed or, better yet, I should have appealed
to him through some influential intermediary. Such was our chief.
Actually, from this moment on he became even cooler toward me.

Affairs at home were no better. Life in the home of the Engel'-
gardts was not to my liking. My situation there seemed to be awkward, and I spoke from time to time to Alexander about taking a
separate apartment and living by ourselves. However, my brother
was at that time on such friendly and close terms with the Engel'gardt family, particularly with Sofia Grigor'evna, that he would
not consider the idea of parting from them. As a matter of fact,
he seldom went anywhere and spent most of his time at home.
In his department he discharged his daily duties in an exemplary
fashion. He was at the time the desk head in the Cadastral Division
of the Third Department of the Ministry of State Property. I too
went regularly to my office and returned only for dinner. Imitating
my brother, I also fell into the habit of staying at home in the
evenings and only seldom went to the theater or to see one of
my relatives. Paul Vasil'evich Engel'gardt frequently played cards
with his usual partners—Ushakov, Melin, and his brother-in-law,
Baron Nolde, when he was in Petersburg. In addition, my brother
and I, together with the kind Colonel Tulubaev, would make up a
party of whist with Sofia Grigor'evna. Sometimes the dependent
in the home of the Engel'gardts, an old and unique spinster, Irina

Ivanovna Travshinskii, the daughter of some priest, would join us in a card game. If I mention also one of our acquaintances from Radzivilov, Attse's wife, I think that I would not omit a single name from the usual members of the Engel'gardt's circle. I must confess that for a young man of my age this society presented little of interest. Thus time passed until the summer of 1840 when I decided to leave the Engel'gardts' home, but before telling of the circumstances under which I did, I should like to say a few words to honor the memory of my kind friend Jean, whom we lost a few months earlier.

In the spring of 1840 he was stricken by a nervous fever which lasted for some time and which brought him to his grave. His last moments were terrible. In his delirium he repeated the words, "Execution, scaffold!" One could judge by the expression of his face that he suffered agonies. I hardly left his bedside, and for a long time after his death I was under the influence of the suffering he experienced in his last moments. Jean Shul'ts was hardly twenty-four years old. He was full of life and nothing foreshadowed such a premature end. In him I lost a friend and a kind counselor who, having more experience in life than I, took a genuine interest in me and tried to be helpful by warning me of the mistakes of youth. I mourned him sincerely.

Shortly after Jean's death we moved to the dacha. Following my custom, which I preserve to this day, I used to take long walks daily. Once on my return home when passing the garden house which was at the end of the garden, I noticed that Sofia Grigor'evna was sitting there with her husband's nephew, a hussar officer, Vasilii Vasil'evich Engel'gardt, with whom she was intimate. Not wishing to be in their way, I pretended that I did not notice them and passed by. I do not know why, but Sofia Grigor'evna imagined that I was watching her, and she was not ashamed to make broad and obvious hints about it that same evening at tea time and in the presence of her admirer. This infuriated me to such an extent that I rose from the table, went to my room, packed my belongings and sent for a cab. I told my brother that I was going to the city where I would look for a place to live.

My decision which I think to this day does me honor surprised Alexander. He justly reasoned that he too should not be a free boarder in the Engel'gardts' home and followed my example, but he behaved much more decently. He had a talk with his hosts and bid them goodbye before leaving. I confess that I did not have the courage to do so. Moreover, I didn't think that I needed to after the impudence toward me described above.

I never met the Engel'gardts after that.* However, I did think of them with affection at times. I had grown to like their children —four charming boys, particularly the kind Grisha, and I was greatly saddened when I heard of the premature death of this fine little boy. I do not know just what happened to the rest of them. Rumor had it that they completely dissipated their father's inheritance. Sofia Grigor'evna lived permanently in the home of her nephew, Vasilii Vasil'evich, and survived him. She must now [1874] be more than sixty years of age.

At first, my brother and I settled down in the apartment of Paul Dmitrievich Diagilev, who was then chief of a division in the same department where my brother Alexander served. This was a spacious apartment let by the treasury, and it was in the same house where the department was, on the corner of Bol'shaia Morskaia and Voznenskii Avenue.† The Diagilevs were then at their dacha. However, we did not take advantage of their hospitality too long because we soon found a charming and comfortable apartment of three rooms, not counting the vestibule and kitchen, on Bol'shaia Morskaia in the house of Lauffert. We paid for it, if I am not mistaken, about 650 or 700 rubles in bank notes a year, that is, according to the present rate, not quite 200 rubles! I am sure that it must cost three or four times as much now.

I cannot express the pleasure I felt when I realized that I was not dependent upon anyone and was my own master. My brother Alexander and I lived in a truly brotherly fashion and concealed nothing from each other. As my older brother he had priority, of course, but he showed it only in his good influence and councils for which I was always thankful. We arranged our little household

* Not even the hussar Vasilii Vasil'evich who later married Countess Elena Sollogub, a cousin of my wife.

† The magnificent houses of the Ministry of State Property near the Sinyi Bridge had not yet been erected.

beautifully, and in spite of our limited means we lived, one could say, in clover. Since our salaries did not cover all our expenses and since we had surrendered the very small inheritance our parents left in favor of our sisters, we decided to do some literary work which we did fairly successfully. Alexander, who was intelligent and well educated, wrote articles on political economy and statistics for various journals. And I did translations and editorial work for the same journals. Once accustomed to the work, we did not find it burdensome. As a matter of fact, it did not consume more than two or three hours a day so that there was enough time for work and all sorts of recreation. During the holidays our younger and cheerful brother Fedia used to come from the School of Jurisprudence. Sometimes we used to invite our old schoolmates, friends, and relatives and the time passed pleasantly. As for high society, the so-called *grand monde*, we did not think of it then; we had other things on our minds and we limited ourselves to our close circle.

More than anyone else we visited the kind Sul'menevs, of course, where we felt completely at home. They spent the summer of 1840 in Novaia Derevnia.

At that time a brother-in-law of Diagilev,* Peter Pavlovich Proteikinskii, a provincial of the old school, lived in Petersburg. It seems that he was always trying to find a good post. I do not know why he could never stay long enough in one place, but he didn't and for this reason he used to come to the capital frequently in search of a new position. I got along with him well because I used to like to listen to his stories and anecdotes about provincial life, and he told them in a masterly way. Moreover, he knew Russia well, having traveled extensively through it. In spite of his corpulence which, incidentally, did not seem ugly because of his enormous height, Peter Pavlovich was a great lover of walks, and wandered daily through the streets for hours. He was a veritable *flâneur*. Once, it was July 1, he dropped in to see me and proposed that we go together to Peterhof for the celebration. I agreed and we went out to look for a cab with two horses. Because of

* I have already said that he was married to my cousin Anna Ivanovna Sul'menev.

the holiday the cab driver, as was the custom, asked a fabulous price. Hard as we bargained, it was still much too expensive. We couldn't take the boat because the last one had already departed and we didn't want to give up the outing. To resolve the difficulty the calculating Proteikinskii suggested that we hire a cab only one way and return from Peterhof by boat. No sooner said than done. We made twenty-seven versts in less than two hours and arrived when the celebration was in full swing. All night long we wandered through the garden, admiring the magnificent celebration which I have had occasion to describe before in these memoirs. Finally, completely exhausted we went to the pier to board the boat. But it was not as simple as that. Every seat was taken. There were only two or three boats, and an enormous crowd of people dragged themselves to Petersburg on foot! We followed them. Our efforts to find a cab or a room in some lodge where we could rest were in vain. Following the mob, we thus walked all the way and only near the gate of Petersburg itself did we find a horse cab to return home. Exhausted beyond words I took off my clothes and stretched out on my bed. I naturally slept the whole day through! We laughed a great deal with Proteikinskii in recalling this forced walk, but at the time, of course, it was not funny.

Shortly after that I took a short trip to Finland with a friend of mine from the Lyceum, Galitskii, whose name was also Peter Pavlovich. The purpose of this trip was to see the magnificent waterfall of Imatra. Our trip was a success. The weather favored us. We took a coach to Pargalov and from there rode in a Finnish cariole—a one-horse carriage which rolled over the smooth road with unbelievable speed. One wonders at the endurance and speed of the little Finnish horses. By nightfall we were in Vyborg where we stopped in a decent and clean tavern. For a long time I could not fall asleep and listened to the doleful singing of the night watchmen who walked through the streets calling out from time to time to the people to rest. Someone told me that this call could be translated into Russian as follows: "The midnight strikes, the hour is late. Sleep. God does not sleep for you." But whatever its meaning, the monotonous singing in the middle of the general quiet has a certain charm and unconsciously induces meditation. These watchmen guard the city not against thieves, who are practically unknown in honest Finland, but against fires, reminding the people to put out the fire everywhere.

In the morning Galitskii and I went sightseeing in the city. It is situated on a hilly site which makes walking through the streets tiring. The city offered nothing exciting, but one is pleasantly surprised by its modest, primitive, and neat appearance. On returning to the tavern for dinner we found in the dining room a small group of amiable and jovial officers from a Finnish garrison and a traveling German artist. We immediately became acquainted with them. After dinner the German artist took us to his room to show us his paintings. They were all portraits in oil. Galitskii expressed a desire to have his portrait painted and the artist promised to have it finished in three days. Moreover, he did so although the portrait was of doubtful quality.

In the meantime one of the obliging Finnish officers (I cannot recall his name) took upon himself the arrangements for our trip to Imatra, and the following day we set out in a post cariole. I do not recall the exact distance from Vyborg to Imatra, but I do remember that we returned to Vyborg the following day. Our road led through a magnificent primeval region with many mountains, cliffs, waterfalls, and lakes. Seven versts before we reached Imatra we could hear the roaring of the famous waterfall, and when we approached it, the roar drowned out our own voices. The water does not fall from a very great height, but it has a great number of rapids. The foam from these rises very high and scatters in the air in a silvery spray. To show us the force of the waterfall our coachman chopped down a large spruce tree and threw it into the water. In one instant it turned into splinters. We then tasted some freshly caught trout at the nearby tavern and found them worthy of their fame. On our way back we were caught in a storm. Lightening and claps of thunder among the mountains and cliffs produced a remarkably majestic effect.

From Vyborg we took another trip to *Mon Repos*, a magnificent estate not far from the city, situated on the shore of the sea and belonging to Baron Nikolai, our former ambassador to Copenhagen. Beautiful though it was, it seemed austere and sad to me. I, of course, would not choose such a retreat even after such an uneasy life as that led by Baron Nikolai who participated in all the important political events of the opening of the present century. I well understand why none of his sons, each one of whom had a brilliant career, would dream of following the example of their father. If I am not mistaken, *Mon Repos* passed into someone else's hands.

We then returned with Galitskii to Petersburg by the same means. We were pleased with our interesting trip which had lasted about a week.

I forgot to mention that at the beginning of the same summer, namely, in the early part of June, 1840, I accompanied Uncle Fedor Petrovich and his wife Iulia Vasil'evna to Kronstadt. They were sailing abroad. At that time Fedor Petrovich began to be unwell. However, despite his poor health he continued with the same zeal his difficult assignment to the grand duke, whom he practically never left. On noticing his condition, the tsar not only suggested but ordered him to go abroad for a cure. The trip, however, did not prove helpful to Fedor Petrovich. He returned almost in the same state of health, suffering as before from sharp spasms in his stomach. None of the famous physicians could cure him. Finally on the advice of his sister Natalia Petrovna he decided to consult the homeopath Oks, who was the Sul'menevs' family physician. In a relatively short time Oks cured Fedor Petrovich radically, and only thirty years after that did he feel the recurrence of the same ailment, of which he was soon cured again by resorting again to the homeopathic remedy of Oks. This circumstance is well known to me, because my uncle had the attack while he visited us in Switzerland in 1869.

Although not in the whirl of high society, I spent the winter of 1840 to 1841 pleasantly. The Sul'menevs' means were apparently improving all along, and they gave some nice soirées, where we always had a gay time. Among the young people there were many charming and educated young ladies. I liked particularly Liza Vistengauzen, who later died in the prime of life from consumption, her cousin, Miss Gedeonov (she married one Bodisko, a guards officer), but more especially the dreamy Anneta Bulatov. I learned later that her fantasies took on a religious coloring, and she took the veil. The concern and friendship that I found in the family of the Sul'menevs brought me so close to them that I used to come to see them almost daily. I shared with them every joy or sorrow, especially with my kind and wise cousin Katia. I recall with pleasure also my old friends Marfa and Alena Nikolaevna Brown, to whom I am thankful for their sincere and warm interest in all circumstances of my life.*

Fedor Petrovich Litke gave extremely interesting soirées once a week which many celebrated scientists and travelers attended. There I met Behr, Midendorf, the Chikhachev brothers, and many others. Sometimes there were musical soirées with such well-known artists as Maurer and Prune. Only classical music was performed. That year Litke lived in the renovated Winter Palace. While it was being restored after the fire, he occupied the first floor of the house of Menshikov on Nevskii Prospect and Karavannaia. At one of the soirées I met Konstantin Konstantinovich Bazili, who was then vice-counsul in Jaffa (Palestine) and who was in Petersburg in connection with affairs of his post. I shall refer to this interesting person from time to time in my memoirs in connection with my work in the East. For the time being I should like to say that he was by birth a Greek from Constantinople, educated in the Lyceum of Nezhin, and began his civil service as interpreter in the squadron of Admiral Peter Ivanovich Rikord, which was in the Archipelago after the battle of Navarino. After the departure of the squadron, Bazili succeeded in obtaining the post of consular agent in Jaffa. This was his *point de départ* for a brilliant diplomatic career. He must be credited for working hard to educate himself and for becoming an enlightened person. He did not overlook his financial interests either. Without a kopek in his pockets when he began his civil service, he is now enormously wealthy.

During the winter about which I write I frequently attended the Noblemen's Club where magnificent balls and concerts were given. The presence of the tsar's family and the entire court added great brilliance to these celebrations. I have already had occasion to

---

* I must say a good word also for the blind old lady, Maria Andreevna Pozhilov, who also lived with the Sul'menevs. She was a sister of Ekaterina Andreevna Litke (my grandfather's second wife). Maria Andreevna was good at reading cards. She satisfied with pleasure particularly the curiosity of our ladies who wanted her to tell them their fortunes. The poor old lady used to spend entire evenings in this innocent occupation!

speak of the magnificence of our imperial court of that time. All
three of the tsar's daughters were remarkably beautiful, especially
Olga Nikolaevna, who was bewitching. Elena Pavlovna, the wife
of Grand Duke Mikhail Pavlovich, was also in the full bloom of
her beauty. Almost all the principal members of the court were
distinguished by their handsome appearance. Who does not recall the stately Prince Vasilii Vasil'evich Dolgorukii, the celebrated
master of ceremonies Count Vorontsov-Dashkov, and so many
others? Towering over all of them was the majestic figure of the
tsar himself who was rightly considered to be the handsomest man
of his day. Naturally all eyes were directed toward the tsar's family
and the persons surrounding them.

As for my work, I never stopped trying to obtain some appointment abroad. At my request my uncle Fedor Petrovich, who always took the warmest interest in me, had appealed for a year or
more to Count Matvei Vingorskii, whom he knew well through his
relations at the court, to act as my intermediary and to speak to
Seniavin. As director of our Department of Personnel (it was called
then the Department of Economic and Statistical Affairs), Count
Vingorskii * saw Seniavin every day and willingly undertook to
arrange matters for me. Several days later he wrote Litke the following brief note which I have preserved among my papers for a
long time: "*Seniavine dit du bien de votre neveu: il va l'envoyer à
Constantinople.*" My joy was great when I read these lines. I knew
that there was a vacancy in the Constantinople mission for an assistant secretary. Moreover, this was the place that attracted me in
particular. I had no doubt that I would receive it soon, and began
making plans for the trip to the Bosphorus. But how terrible was
my disappointment when I learned that someone else was designated to this post. After a while I decided finally to have a talk
with Seniavin and went to call on him in his dacha on Kamennyi
Island. Lev Grigor'evich received me courteously (which had not
always been his custom toward his employees) and gave the appearance that he really was interested in me. He told me that in a
very short time another vacancy was expected for an assistant

* At the same time he was *Hofmeister* of the grand duchess Maria Nikolaevna who married Duke Maximilian of Leuchtenberg in 1839.

secretary in Constantinople, but that for some reason the official who would be appointed to this position would have to remain some time in Bucharest, attached to the newly appointed consul-general in the Danubian Principalities, Vladimir Pavlovich Titov. "If this is in accord with your views and desires," added Seniavin, "then I would suggest that you go immediately to Titov and ask him to take you with him to Bucharest, but do not mention one word to him about our conversation." I replied to Lev Grigor'evich that I hesitated to do so because I did not know Titov and he might think my request strange, particularly if I were prevented from referring to the recommendation and support of my immediate superior upon whom all appointments depended. Seniavin did not agree with me and kept on insisting that I do as he told me. Whereupon he gave me Titov's address. Titov had just married Countess Elena Mikhailovna Khrebtovich (a relative of Count Nesselrode) * and lived in the house of Vel'tsyn on Malaia Morskaia. I said goodbye to Seniavin, went home, thought the matter over, and decided against going to Titov. I suppose that this angered my haughty chief, because he began to avoid me openly. The person appointed to the mentioned post was one Kurliandtsev who had just returned from China. Other vacancies became available later, but I was never considered for any of them. Thus the fall of 1841 passed. I felt that I was in disgrace and, having lost all hope of ever getting a position abroad, I began to think of transferring to another department. However, before relating how I found a way out of this situation I should like to say something about the last months which I spent in Petersburg.

After the winter amusements everyone awaited the arrival of the bride of the tsar's oldest son, Alexander Nikolaevich. The triumphal entry of the most noble bride, favored by excellent weather, was a great success. The procession moved slowly from the city gate, greeted and accompanied everywhere by an enthusiastic crowd which received the royal betrothed with deafening cries. This was truly a national holiday fully expressive of the sincere love and de-

---

* She was one of the sisters of Count Khrebtovich who was married to Nesselrode's daughter. His other sister, Countess Maria Mikhailovna, was married to the ambassador in Constantinople, Apollinarii Petrovich Butenev.

votion of the Russian people for their tsar. To be sure, one could not help being proud of him. He was magnificent in his dress uniform, mounted on a playful horse, always keeping to the right of the rich carriage occupied by the empress and the young bride, whose charming and modest appearance elicited universal approval. One had to admire the affectionate character of the groom who was also on horseback on the other side of the carriage. The wedding followed at the end of April. I remember particularly the number of celebrations and holidays given on this occasion. Since my position was not as yet high enough to be received at court, I was present only at the ball in the Noblemen's Club whose magnificence surpassed all my expectations. Never have I seen the capital so brilliant. I confess that I spent most of the day on the streets admiring the various public spectacles, the illuminations, and the magnificent parades.

Early that summer Grand Duke Konstantin Nikolaevich made his first trip abroad chaperoned by Litke. I recall that one morning my uncle Fedor Petrovich came to see me and asked me to go to the office of the governor-general to get the passports for the grand duke and his retinue under the name of Prince Romanov. I accepted this errand with pleasure. I put on my full-dress uniform and went to the office. With an important and mysterious air I transmitted Admiral Litke's request. A most respectful reception on the part of the chief of the office followed. He couldn't find a seat good enough for me and at once produced the passports. I could not help thinking that if I had come here as a private person, the reception would have been different. No doubt I would have had to sit and wait a long time before His Excellency would have deigned to listen to me. The room was full of people, many of whom had been waiting for hours.

Around this time I happened to meet two persons to whom fate later brought me very close. Sitting at my desk in the department one day, I noticed two men enter Seniavin's office, one a prominent-looking man of middle age wearing glasses, and the other a young man with a small black mustache. The distinguished bearing, remarkably expressive face, and piercing eyes of the former impressed me just as much as the likeable handsomeness of the latter. They were, as I found out later, Prince Alexander

Mikhailovich Gorchakov, counselor of our embassy in Vienna, and his nephew Prince Lev Egorovich Cantacuzino. They had come to talk to Seniavin about the lawsuit the Princes Cantacuzino were carrying on in connection with their estate, Skito-Hangu in Moldavia, where they owned large properties. Little did I dream that several years hence I would be related to them and that the estate in question would constitute the dowry of my future wife!

Prince Gorchakov was at that time in Russia on leave and thought of leaving his post because of some difficulties with Count Nesselrode. He thus remained in Russia for about two years without occupation. He spent practically all his time in Moscow. Fortunately for Russia fate had a brilliant career in store for him. However, he remained more or less in disfavor for a long time because his appointment as ambassador in Stuttgart, where he was kept for eight years, did not, of course, do justice to his position or his brilliant gifts. The events of 1848 in Germany advanced him somewhat. An intelligent diplomat was needed at the Frankfurt Parliament, and he was sent there as deputy while maintaining his post at Stuttgart. Later, in 1851, our relations with Austria became complicated, and he was made ambassador in Vienna. His work there and the further fortunes of our famous state chancelor are known to everybody.

I can say positively that the appearance of both interesting visitors in the Asiatic Department produced the same favorable impression on my co-workers there as it did on me.

Speaking of the department, I cannot pass in silence another incident which comes to my mind although it happened several months before the one just described.

I usually worked in the department from 12 to 5 o'clock in the afternoon. Once it happened, I forget why, that I returned home much earlier, around 3 o'clock in the afternoon. I was met at the gate by our porter who told me that he did not think that it was fair that we should entrust to a total stranger the job of taking our belongings and moving them to another apartment. We, of course, knew nothing about it. We did plan to move into another larger apartment, but not immediately. I told all this to our porter and asked him where our servant Ivan was. To this he replied: "He hasn't been home for some time, and in the meantime some men have entered your apartment and have packed all of your things in bags. They were ready to take them out, but they met me in the doorway. I asked them why they were taking your things and where. They said that they did so because you had ordered them to, and that you were moving to another apartment. I didn't think all was well so I locked the door and wouldn't let them out. And I am here guarding them." There was no doubt but that the men were thieves. I thanked the porter for his precautions, and together we ran up the stairs to our apartment. On unlocking the door I met two strangers of a repulsive appearance, and one of them of enormous height. They held empty bags in their hands. They repeated the same lie to me, saying that Ivan had hired them to move our things. "But where is Ivan?" I asked. "We don't know where he went. We are waiting for him." I asked the porter and another worker from the next apartment to watch the thieves lest they should run away and began to look through everything to make sure that nothing had been taken. The drawers in the tables were all locked. In the meantime my brother Alexander arrived and greatly alarmed by what I told him, he too began to look to make sure that nothing was missing. Finally, we discovered that we missed the diamond order of St. Anna which had belonged to our late father. To all our questions the thieves repeatedly swore that they had taken nothing from us, demanding that we free them or call the police. Needless to say that by keeping the diamond cross they feared nothing because with the corruption and bribery prevalent among the police authorities they could count on going free. It was obvious that they had hidden the cross well because they showed no uneasiness and were quite brazen, continuing to insist that we call the police. We naturally did not want to resort to this measure until we had found the cross, but all our searches were in vain: We found nothing in their pockets or boots. When I noticed that one of the thieves, who answered our questions with insolence, continued to wear his hat despite our presence I raised my hand to knock it off. I noticed that my gesture made the thief extremely nervous and that he held on to his hat. This movement seemed suspicious to me, and I quickly grabbed his hat and began to examine it. I felt something hard sewed into the lining. We tore the lining and there we found the cross. Only then did we send for a police officer. In the meantime our Ivan finally returned and in despair told us of his unfortunate exploits. It seems that early that morning one of the two men, whom he had met in some tavern, came to see him and invited him to go with them and have something to eat after we had gone to the department. Without suspecting anything Ivan agreed to accompany his new friends, who got him drunk, drove him to the outskirts of the city and, throwing tobacco in his eyes, left him there. All of this was probably true, but, having lost confidence in Ivan, we could not keep him any longer. Because of his irresponsibility we almost lost our possessions.

In place of him we took one of our serfs who had been sent to us several years before from Radzivilov—a strong young boy, Maksim. He had already been in our service but because of a theft which occurred while he was with us (our late father's gold watch was stolen), we had decided to send him away to some tailor to learn the trade. I shall speak about this later.

At the end of May, shortly after I celebrated my coming of age, I became ill. I recall that in the morning while I attended a concert given by some Italian singer in the house of Count Kushelev-Bezborodko, I developed such a headache that I could not stand the sound of the music. Nevertheless I went for dinner to the Sul'menevs who lived then on Pargalov Road in the dacha of Petrov not far from the Vyborg Gate. This was on May 29, the birthday of my aunt Natalia Petrovna, also of my brother Alexander. There were many guests, and during the dinner a military band, the Latvian Guards Regiment, played. I simply could not stand the noise, left the table, and went into one of the far rooms to lie down. They applied lemons to my temples and used other remedies, but nothing helped. The headache was unbearable. In the evening my brother took me home. The following day I felt worse. A doctor was summoned who put me to bed. From that moment on I do not recall anything. I had a severe case of nervous fever. On the eve of my illness I had read the memoirs of Mlle. Cochelin, a lady in waiting to the Empress Josephine, and in my delirium I talked about the court of Napoleon. I regained consciousness after a terrible storm broke over Petersburg. This was on the fifteenth day of my illness, and Doctor Tsinder who took

care of me as if I were related to him announced that there was
hope that I could be saved. All my good relatives visited me. I
recall particularly the visit of my sweet aunt Iulia Vasil'evna who
came to see me from Tsarskoe Selo. I shall never forget the warm
interest they all took in me. When I began to feel better, my
brother Alexander, who would leave me only during the time when
he had to be at work, told me that it always worried him when he
had to leave me in charge of Maksim alone because this little boy
was terribly afraid of me during the moments of my attacks. Once
on his return home my brother heard awful screaming coming from
my room. He hurried there to find Maksim lying under me on the
floor and, frightened to death, shouting for help. In my delirium
I often tried to get out of bed, and Maksim had strict orders always to hold me back. This time my tall and lean figure frightened
him so that he wanted to run away, but when he felt my grip his
legs gave way under him and we both dropped to the floor. I lost
consciousness and Maksim began to scream desperately thinking
that he was lying under a corpse. God knows how long we would
have remained in this position had not my brother fortunately arrived. I should mention that Maksim was then already a strong
seventeen-year-old boy!

The kind Doctor Tsinder rejoiced in my recovery no less than
did the rest of my friends and relatives. Of the ten patients stricken
with nervous fever whom he was treating at the time I was, he said,
in the most critical condition. But thanks to God I was among the
few he was able to save. I hear that he himself is no longer among
the living although he was then quite young. I recall this man with
gratitude.

As soon as I was able to leave the house, I went to the department to ask Seniavin's permission to spend a few weeks in the
country to restore my health. On seeing me looking thin and weak,
Seniavin was terrified. He rushed toward me to hold me up and
told me not to worry about my work until I was completely recovered. I took the occasion to thank him for raising my rank, during
my illness, to that of assistant clerk which gave me a considerable
increase in salary. This increase was useful because with the introduction of the new monetary system around this time everything
in Petersburg became expensive. Up to that time all financial

negotiations were conducted with bank notes. A bank note ruble was equivalent to a franc, and a kopek to a centime. The sudden increase in the monetary unit from the paper ruble to the silver ruble caused a fast rise of prices on all commodities because the merchants, storekeepers, and industrialists generally could not, or did not wish to, understand that the new rubles and kopeks were worth almost four times more than the others. It seems to me that the wise minister of finance Count Kankrin made a considerable blunder in this respect. The people who suffered from it most were the government employees and persons who lived on the interest from their capital, the so-called *rentiers*. As a German, Count Kankrin thought it useful, of course, to put the ruble on a par with the *Thaler*, but in Germany the people were accustomed to using the *Silbergroschen*. In our country the reckoning in kopeks (in spite of the fact that its value was increased) was still considered trivial, although in reality it was different. I am convinced that it would have been to our common advantage to return to the old system, that is, to put our ruble on a par with the franc.

After securing Seniavin's permission I lost no time and went at once to the Sul'menevs where I spent about three weeks. The peaceful and pleasant life among my close friends, the pure air, and the walks in the pine grove which surrounded the dacha, where I spent many hours, fully restored my health.

The Bykovs also lived in one of the dachas of Petrov, of which there were four, I think, and the Diagilevs lived not far from them near the Forestry Corps where new dachas were just being erected. Thus I was in the pleasant company of friends and relatives.

On my return to the city I resumed my duties in the office. Once while sorting some papers I came across a communication addressed to the consulate in Moldavia requesting that permission to remain in Petersburg be granted to the Collegiate Assessor August Ivanovich Ebergardt, our agent at the Skuliany quarantine, who also performed the duties of secretary at the consulate there. Immediately it occurred to me that Seniavin intended me for this position; I became even more convinced of this when I learned from Konstantin Ivanovich Lents, a co-worker occupying the desk next to mine, that Ebergardt who was his relative did not intend to return to Moldavia. From that moment on I began to consider what to do if I should be offered this position. The work in the consulate did not satisfy my ambition. On the other hand, to reject an offer by Seniavin who had become even haughtier after his appointment as director would mean to fall completely into disgrace and consequently to lose all hope for the career of my choice. After discussing all this with I. D. Ginkulov I decided on his advice that if the offer was made to submit to the will of the director but with one qualification: to ask him to keep me in mind should a vacancy occur in one of our missions in the East. My brother Alexander approved this decision, reasoning as I did that the main thing was to leave the department.

Several days after this my fate was decided. All the circumstances of the event, so important in my life because of its consequences, remain in my memory to the minutest detail. I was checking a dispatch which had to be verified by K. K. Zlobin, the head of the Greek desk. This was in the room where the magnificent full-size portrait of Alexander I hung. Suddenly we heard the director's bell, followed by the quick steps of a messenger who was coming toward our room. I interrupted the reading and told Zlobin that I was sure that Seniavin was ordering me to appear. My premonition was correct. I confess that my heart beat fast when the messenger addressed me by saying: "Please report to the director!" How many appointments have I had after this one during my long career, and much more important ones than this, but not one of them had such a tremendous effect on me as this first one to Moldavia. I sensed that the foundation was being laid here for my entire future. My excitement was not against me, rather it was in my favor. When I calmed down a bit I entered the office. I listened to the director's offer, to which I replied that I was glad to accept Ebergardt's place, but that I hoped that this would not disqualify me in the future from being appointed to one of our missions. Apparently satisfied with this reply, Lev Grigor'evich assured me that I would not remain long in Moldavia and would be transferred to Constantinople. He then outlined my duties and added that until Ebergardt received another post I would be getting the allowance of the second dragoman of the consulate whose post remained vacant. This did not please me too much because the salary of the second dragoman was only 800 rubles in silver or even less. But with the hope of better things to come in the future I agreed to this also. My formal appointment followed on October 19, 1841, and I began to make my preparations for the trip.

Thus I served in the Asiatic Department exactly three years. Although I had acquired a degree of knowledge, and experience in affairs and with people, the service nevertheless had been without any benefit for my career because I was not given the slightest reward and in the end received a post abroad which any beginning official could obtain. Nevertheless I was glad to leave Petersburg where I was so unlucky, and to see the world.

Nicholas Karlovich Giers in the 1870's

V

MY FIRST TRIP TO MOLDAVIA
(September–October, 1841)

MY PREPARATIONS for the long journey did not take too much time. I bought a comfortable carriage and took as my traveling companion Maksim, whom my brother Alexander surrendered to me. After saying farewell to my friends and relatives I spent the last few hours alone with my brothers. Exactly at midnight the post horses arrived. Parting from my brothers was hard, and I could not even guess how long our separation would last. I left in the last days of September, but I cannot recall on exactly what day. The night was dark and damp. We drove slowly through the city to the gate near the Arch of Triumph. From here we took the highway, but we could use it only for several stations because we had to turn on to the Belorussian Highway where the road proved to be extremely bad. The horses could barely drag the carriage over the terrible mud, and we went at a slow pace practically all the time. Moreover, the traffic over this road was so great that we had to wait for hours at practically every station because of the shortage of horses. If we could only have used these long stops to rest. But this was impossible. Because of the great number of travelers it was impossible to find a corner in the station. My traveling credential which read *on government business* was not of much use to me because among the travelers were such important dignitaries as General Kiselev, Senator Count Zavodovskii, and others who, of course, had to be given preference over me. I became impatient, but hearing the scolding the station masters received from the more important travelers because of the shortage of horses, which of course was no fault of theirs, I calmed down. So it continued for several days. After Velikii Luki the road improved, the number of travelers decreased, and we raced. I felt better, talked to Maksim, and was happy in the anticipation of seeing my sister Valeria soon. She, of course, had been waiting for me for a long time in Zhitomir. In a moment of such pleasure I composed the following couplet which I kept singing:

> At Velikii Luki station
> Ended all my tribulation

In fact, from then on all went well. The sun came out, and we had clear warm weather to the end of our journey.

Feeling the need of at least a brief rest, I stopped over night in the city of Vitebsk in some fairly decent tavern. The city had a good appearance in spite of a great number of Jews. From early in the morning we were besieged by petty commissioners (Jewish commissioners) with offers of all sorts of services. You simply could not get rid of them. The local inhabitants, particularly the Poles, are so accustomed to them that they do not take a step without them. They serve them as intermediaries in all of their affairs.

After Vitebsk I made another overnight stop in Mogilev, and I rode from there without stopping to Zhitomir where the Anferovs were happy to see me because they had begun to worry, wondering about the cause of my delay. On the second or third day my sister Emilia Speranskii came there from Berdichev.

My first duty on my arrival in Zhitomir was to visit the grave of my dear mother who is buried there in the Russian Orthodox cemetery. With sadness and tender emotion I recalled how impatiently she had awaited spring to go to Petersburg to establish residence there after my graduation from the Lyceum. My sister Valeria, who was with my mother during the last moments of her life, told me that in her delirium she often spoke of me. Her last thoughts were of her children. How could we help but treasure her memory! I prayed from the bottom of my heart over her remains, sacred to me. The cast-iron monument with a bronze cross over her grave is simple but proper.

Nikifor Ivanovich Anferov, whom I had had occasion to meet in Petersburg when he came there on business, held at that time the post of provincial procurator in Zhitomir where he was liked and respected. Through him and my sister I immediately met all the local officials, beginning with the governor, General Loshkarev, and also some of the Polish landowners, among whom I recall Count

Tishkevich. I received invitations from all sides either to dinner or a soirée so that I had a gay time during my stay of about a week in Zhitomir.

Life in the provinces seemed to me much pleasanter than in the capital because at least one could count on a hospitable reception everywhere. There were many young Polish people at the party in Loshkarev's home, and I admired the relaxed grace with which they danced the mazurka.

The Anferovs had then two or three children. I recall only one little girl, Anniuta, who amused me with her childish chatter.

From Zhitomir I went together with my sister Emilia to Berdichev where I remained three days. Her husband, Ivan Petrovich Speranskii, lived there as an official on special mission from the Ministry of Finance to guard against smuggling, widespread throughout our entire western border because of the Jewish population in that territory.

We made this trip in only a few hours. I recall that while driving through the forest we overtook a convoy escorting several prisoners. They were robbers who had been caught recently. Riding through the forest at night in those regions then was considered dangerous. It is said that robbers seldom attacked travelers who used post horses, but for the most part merchants who did not use relay horses, particularly the Jews.

Berdichev is a large, rich, and thickly populated city, inhabited almost entirely by Jews, which gives it quite a dirty appearance. It is hard to imagine anything more repulsive than this population in their hideous medieval costumes, with their long locks of hair or *peisy*, always uncombed, in their short trousers (*culottes*), their socks always dirty, with torn shoes, and ugly wide-brimmed hats. A Jew presents the most comic and pitiful sight. Berdichev conducts an extensive trade in all foreign-manufactured goods; it has magnificent stores where one can find almost anything, and the Polish landowners from the surrounding area come there to shop.

The house which the Speranskiis occupied was situated, thank God, far from the city and consequently far from the dirty Jews. I spent my time in their home entirely in the family circle which, of course, I found very pleasant. Of the children they had then I recall the lovely Masha and the handsome Sasha, who sadly ended his life on graduating from Petersburg university. He was an extremely intelligent and gifted little boy. Unfortunately, he became infected with the malignant spirit so widely prevalent among our youth around 1860 and became completely confused. Arrested with a few other of his university friends for some conspiracy, he wanted to hang himself, became melancholy, and soon died!

In the home of the Speranskiis I was surprised to meet our old nurse Mrs. Anna Sosnowska who was in their service. I recognized her immediately although I had not seen her since I was eight years old.

I said goodbye to the Speranskiis and went directly from Berdichev to Kamenets-Podol'sk where other relatives close to my heart were waiting for me.

I covered this distance with remarkable speed. The post horses along all this line were so large and strong that they could have been harnessed to any carriage in the capital. The road throughout is good, and in some parts passes through picturesque regions. The people too looked better than those I had seen in the country I had passed before. They obviously live better, and that stands to reason. The soil in the province of Podolia is remarkably fertile and the climate excellent. It also pleased me greatly that the memory of my kind father was sacredly preserved both in Volynia and Podolia. In almost every station, when I registered my credentials the station master would come out and ask me if I were the son of the late Karl Karlovich. At my affirmative answer I invariably heard: "There was a kind man! He was our benefactor! May he rest in peace!" Then inquiries about the family would follow. Some of the station masters knew me as a child. It did my heart good to hear these genuine, sincere expressions of love and devotion to my father.

Among the stations which I passed I recall particularly Proskurov, Old and New Konstantinov, Bykhov, and especially Tynaia, situated on the magnificent estate of Butiagin, the father of Eudoxie Butiagin, a pupil at the Catherine Institute, about whom I spoke earlier in these memoirs, and whom I had met in the home of my aunt Baroness Elisaveta Petrovna Rozen in 1831.

In my opinion there is no other place in all of Russia as enchanting as Podolia, and if I were looking for a haven in my old age, I would wish nothing better than an opportunity to acquire even a small estate in this blessed territory and to settle there with my dear little family. Although it is far from Petersburg where, no doubt, most of our children would have to live, this distance can be covered now in no more than three days and in even less time because of the railroads.

In Kamenets-Podol'sk my uncle Fedor Karlovich received me with open arms. My sister Iulia whom I had not seen since childhood, from the time I left Radzivilov to be exact, also lived there. She was married to Colonel Nicholas Ivanovich Kolonov, commander of the battalion of the Home Guard in Kamenets-Podol'sk. My uncle Fedor Karlovich had then only one daughter, Valeria, a lovely twelve-year-old little girl whom he adored. His sister and my aunt Anna Karlovna Maiet * lived in his home also. In this family circle I spent the most pleasant five or six days.

Fedor Karlovich lived in luxury and was on friendly terms not only with the Russian government officials but also with the Polish nobility, among whom he commanded great respect. The best society of this region used to gather in his hospitable home. Naturally under the circumstances he never wanted to leave Kamenets. He was extremely kind to me, drove me daily to show me the outskirts of the city, reminisced about his past with me, and told about life in Paris after it was occupied by the troops of the allies in 1815. He also expressed his views on the French literature of the 'thirties and bewailed its present trend so detrimental to the younger generation. Fedor Karlovich knew some of the landowners of adjoining Bessarabia who had estates in Moldavia and who used to drop in to see him in Kamenets. He praised particularly the Princes Cantacuzino and the Krupenskiis, and advised

---

* Her sons, Fedor and Konstantin Petrovich Maiet, live in Petersburg. Fedor Petrovich is a retired seaman, a bachelor. He spends most of his time in the club. Konstantin Petrovich, who served in the post office department and enjoyed the particular favor of his superior, Prince Alexander Nikolaevich Golitsyn, was married to a French woman, Mlle. Constance (I do not recall her last name), who died, if I am not mistaken, in 1839. About eight years after that Konstantin Petrovich married for a second time, our mutual cousin Valeria Fedorovna Giers, whom I mentioned in these memoirs.

me to cultivate them. This advice, as we shall see later, I followed and to an extent which surpassed his expectations.

While in Kamenets I naturally frequently visited my sister Iulia. Her husband, a brave and honored military man, decorated with the Cross of St. George, loved to tell of his campaigns, and he knew how to add life and interest to his stories. To be sure, it was terrible to think what privations and dangers our army had to endure during the Turkish campaign of 1828, particularly in the Balkans, and how many people fell victim to the plague from which the troops suffered much more than from the enemy's fire.

On leaving Kamenets we raced straight for Moldavia. My course was in the direction of Khotin where the old Turkish fortress, so famous in the history of our wars with Turkey, was preserved. I crossed the Dniester and found myself in Bessarabia.

This country presents a totally different aspect from the province of Podolia. There are practically no forests. Moreover, prosperous villages with churches and the large estates with parks of the gentry are more seldom encountered. However, the soil is just as fertile as in Podolia, and magnificent fields of wheat and particularly of maize stretch out for many versts. At the time of this trip the harvest had already been collected, and the fields, surrounded by bare hills, presented a sad appearance. For a distance from the Dniester one could still hear Russian speech, Ukrainian to be more exact, but little by little it disappeared and was finally replaced by the Moldavian tongue. This part of Bessarabia has remained outwardly practically unchanged from the time of its annexation to Russia.* The native, or the "ţăran" [peasant] as he is called there, hardly understands Russian. He preserves his native costume, resembling that of the Ukraine. He wears a shirt, completely open at the neck and exposing his hairy chest, with a wide belt which serves also to hold a knife. In the summer he wears a hat with an unusually wide brim and in the winter a sheepskin cap. The approach of Moldavia could be felt with every step. Finally (that was on the second day after my departure from Kamenets), I approached the Prut and stopped at Skuliany in front of a Russian customs house. There remained but to cross this narrow river in order to step on to Moldavian soil.

* This annexation followed in 1812 according to the Peace of Bucharest.

PART TWO

Service in Moldavia

(1841–1847)

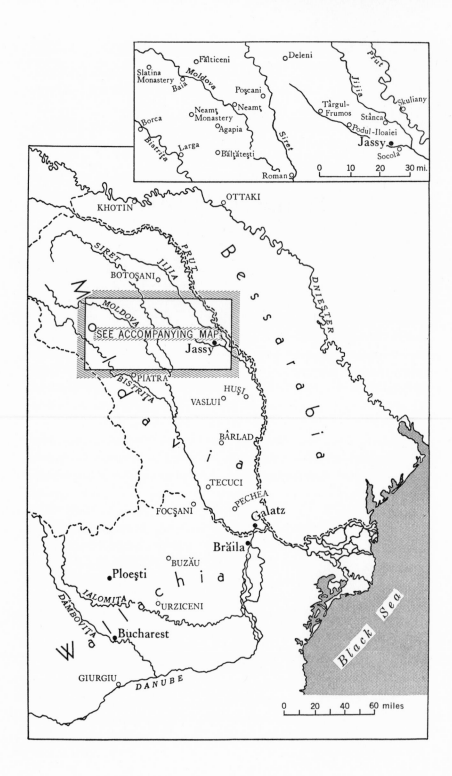

Inset map (upper)

Fălticeni

Deleni

Slatina
Monastery

Moldova

Baia

Poşcani

Târgul-
Frumos

Prut

Jijia

Skuliany

Neamţ
Monastery

Neamţ

Stânca

Borca

Agapia

Podul-Iloaiei

Jassy

Bistriţa

Larga

Bălţăteşti

Siret

Socolă

Roman

0   10   20   30 mi.

Main map

OTTAKI

KHOTIN

SIRET

JIJIA

PRUT

Bessarabia

BOTOŞANI

MOLDOVA

DNIESTER

SEE ACCOMPANYING MAP

Jassy

PIATRA

BISTRIŢA

Moldavia

HUŞI

VASLUI

BÂRLAD

TECUCI

PECHEA

FOCŞANI

Galatz

Brăila

BUZĂU

Ploeşti

IALOMIŢA

Wallachia

URZICENI

DÂMBOVIŢA

Bucharest

Black Sea

GIURGIU

DANUBE

0   20   40   60 miles

EDITORS' INTRODUCTION TO PART TWO

ON CROSSING THE PRUT, Giers entered a country which was in an ambiguous political position. Although nominally under the suzerainty of the Porte, Moldavia as well as Wallachia had become Russian protectorates following the Convention of Akkerman in 1827. Henceforth, until her defeat in the Crimean War, Russia rather than Turkey determined the course of the internal developments of the Danubian Principalities. After the Russo-Turkish War of 1828 and the conclusion of the Treaty of Adrianople, a Russian army of occupation remained in the provinces, and the territory was given its first modern administrative system under the governor general, P. D. Kiselev.

The Organic Statutes, drawn up for each principality under Russian sponsorship, provided the framework for a highly conservative political system, but one which was, nevertheless, more liberal than that prevailing in Russia at the time. A legislative assembly was created composed of representatives from the boyar class only, and the executive power was to be exercised by a hospodar chosen exclusively from this same group. The hospodars and the assemblies had no direct veto over each other, but each could appeal directly to Turkey or Russia. In practice the Organic Statutes thus created in each principality three separate centers of political power—the hospodars, the assemblies, and the Russian consuls to whom disgruntled political elements could and did appeal. With this organization the Russian protectorate was characterized by bickering between rival political blocs and between individuals anxious for power. In these petty intrigues and minor controversies the Russian consuls often interfered, and because of the overwhelming military and diplomatic power of the Russian government, they were usually able to enforce the acceptance of their views.

The position of the Russian consuls at both Jassy and Bucharest was, nevertheless, not enviable. Because of the extreme corruption and irresponsibility which characterized Rumanian national life in this period, it was indeed difficult for any agent assigned to the principalities to remain outside of the tortuous intrigues and petty conspiracies and still maintain the influence of his country. During the time when Giers was assigned to Moldavia, the Russian government wished to uphold its prestige as the protectorate power, but it did not wish to extend its control or to annex the country outright. The Russian consuls were thus faced with the necessity of enforcing Russian desires through individual acts of legislation without precipitating a real crisis.

After the evacuation of the principalities by the Russian armies in 1834, the principal Russian agent was Baron P. I. Rikman (Rukmann), the consul at Bucharest. The same position at Jassy was held first by Besak, who was followed by K. E. Kotsebu, under whom Giers served. All three of these representatives, as well as Giers, had German backgrounds. From the time of his arrival in the principalities Rikman felt it his duty to interfere actively in Rumanian political life. He was even able to obtain the replacement of the Wallachian hospodar, Alexander Ghica, with the more docile George Bibescu. In Jassy Besak met with a more formidable opponent in the person of the hospodar Mihail Sturdza. Clever, corrupt, and greedy, Sturdza was able to block the attempted interference by the Russian consul. When Besak intrigued with the opposition boyars, Sturdza complained to the Russian chancellor Nesselrode and obtained the recall of the consul. However, when Kotsebu continued Besak's policies, Sturdza was unable to secure the new consul's removal.

The description which Giers presents of the principalities in the 1840's is interesting not so much because of his political comments —he does not attempt an analysis of the government or its policies —but because of his picture of the life led by the ruling boyars under the Russian protectorate. Giers became part of the society he described; he eventually married into the Cantacuzino family whose fortunes are so closely followed in these pages. Although he did not approve the scandals and misdeeds he records, he was always sympathetic with the cosmopolitan and opportunistic society whose amusements and entertainments he so enjoyed.

The social system reflected in the Giers memoirs was the direct outgrowth of the system of government established in the principalities under Ottoman rule. Because of the failure of the Ottoman officials to master foreign languages, they were forced to rely on their subject nationalities for administrative and military duties. For these purposes the Porte regularly employed the Greeks living in Constantinople in high government positions. Since most of them lived in the Phanar, or Lighthouse, district in the capital city, as a group they were known as the Phanariotes. The highest offices open to them were those of grand dragoman and hospodar of Moldavia and Wallachia. Because of the wealth which could be won in this connection, the Phanariotes fought among themselves for these coveted positions with intense rivalry and extreme bitterness.[1] Since they were forced to pay a high price to obtain the office, and since it was necessary for them to send frequent contributions to Constantinople to remain in power, the Phanariote rulers of the principalities were chiefly concerned with milking their province of the last drop of portable wealth obtainable. Nevertheless, despite a series of rapacious governors, the principalities remained a rich prize throughout the nineteenth century, and they were referred to as the "Peru of the Greeks."

However, for the political future of the principalities, the social system inaugurated under this rule was far more significant than the immediate economic effect of Phanariote corruption. The Greeks in their dealings with the local inhabitants duplicated in tone and performance the attitude which their Moslem overlords adopted toward the subject Christians. Contemporary travelers in the principalities were appalled by the atmosphere of the courts and the "pure depotism exercised by a Greek prince who is himself, at the same time, an abject slave."[2] The Phanariote princes were overbearing and arrogant toward their subordinates. To make their own fortunes and to meet the payments to Constantinople, they sold the offices under their control and exacted extraordinary taxes and contributions to the fullest extent of their power. Corruption, initiated at the top, extended down to the lowest levels of administration. Since all offices were sold, the holder of any position tried to recoup his losses from those below him. Moreover, even among the few most powerful families, no common accepted standards of conduct existed. The maladministration of the Cantacuzino estates by the Sturdzas, which Giers describes, was an example of, and not an exception to, the rules of behavior among the great boyars in the principalities.

In 1822, after Ipsilanti's unsuccessful revolt in the cause of Greek nationalism, the Greeks lost their monopoly of the hospodarships, but as a nationality retained their unique position in the principalities. In the period under consideration, they were as a rule sympathetic to Russia because of the bonds of the Orthodox Church, a common Byzantine heritage, and the interest which the Russian government had shown in Greek nationalist endeavors. Under Alexander I Greek nationals, such as Capodistrias, who was foreign minister, Radofinikin, who directed the Asiatic Department, and numerous others serving in the Russian army, had distinguished careers in the Russian service. Close ties also linked many Russian, Greek, and Moldavian families who held estates scattered throughout southern Russia, Bessarabia, and the principalities. In these memoirs it will be noted that most close associates of Giers were of Greek extraction. Giers himself was struck by the non-national character of the ruling classes.

Throughout the 'forties a tide of resentment against Russian rule and against all that Russia represented rose in the principalities. Led by the sons of the boyars who were sent west for their education, this movement eventually led to the revolutions of 1848. In these memoirs Giers makes no mention of, and is obviously unaware of, the growth of liberal and national sentiment which was to culminate finally in the unification of the principalities and the destruction of Russian influence in the country. The Russian protectorate could only be maintained while political power in the principalities remained the monopoly of a few great families. It is their way of life and their attitude toward their country which Giers describes so well.

VI

JASSY (End of 1841)

AT THE PRESENT TIME a trip to foreign lands is one of the most usual events in the life of a Russian. However, in the days about which I write this event was so momentous an occasion, especially for a young man, it is no wonder it fired one's imagination. Therefore, my heart beat fast as I approached the Moldavian border. I expected that I would see a new world beyond this boundary. I also envisioned the commencement of my diplomatic career there in the most attractive, bright colors despite my modest official position.

Our custom's officials at Skuliany treated me well, and I was allowed to pass through the gate almost without any delay. Here I met our quarantine officials with whom I immediately established friendly relations thanks to the thoughtfulness and hospitality of the Inspector of the Quarantine for the Prut, Alexander Grigor'ievich Rozhalin, a kind and venerable old man who had held this post for almost twenty years. Irrespective of the work that accumulated in the consulate at Jassy, it was to be my duty to go regularly to Skuliany when mail from Constantinople arrived in order to be present when the packages were opened and fumigated in the quarantine, lest someone should decide to read the dispatches of the mission and the consulates. Therefore, Rozhalin immediately made me familiar with the place and with the procedure in this matter, which was none too difficult. Thereafter, he escorted me to the river where a ferry waited to take me to the Moldavian shore. Here I was met by the head of customs, one Haralambie, an old Greek, wearing the Moldavian national costume with its indispensable accessory, a long pipe, held in his hands. The entire staff of this dignitary consisted of two or three aids, *slujitori*, or gendarmes, dressed on the order of our Cossacks, and a few Jews who collected the customs duty for a fixed fee that they paid to the Moldavians. Needless to say, they did not even touch my baggage since I enjoyed immunity from customs inspection by virtue of my official position.

On my arrival everyone made haste to secure post horses for me as soon as possible. But great was my surprise when I saw the coachmen, or *surugi* as they are known there, driving ahead of them a great number of small and emaciated horses, running completely free in a herd. After examining my carriage, the *surugi* hitched the horses to it, or rather tied them to it, with thin ropes of poor quality—eight horses in pairs, and harnessed one pair behind the other. Two of them were saddled. The one to the left, near the axle, was mounted by a *surugiu* who seemed somewhat older and who straightened out his rope-reigns in order to guide the first team of four horses, and the other *surugiu* handled the front team of four horses in a similar fashion. I protested against such needless harnessing, saying that I had never in my life seen eight horses hitched to so light a carriage. But my protestations were in vain. However, when I realized that all I owed in payment for the trip from the station to Jassy was a trifle more than one ruble in our money, the great number of horses notwithstanding, I stopped complaining.

No sooner were Maksim and I seated in the carriage when the *surugi* began to shout in the most savage voices, "haide" [giddy-ap] while cracking the unusually long whips with short handles. We raced with such speed over the uneven road, which went uphill all the way, that I expected the coach to turn over or to be shattered into fragments any minute. In spite of all our efforts, it was absolutely impossible to moderate the fury of the *surugii*. Without paying the slightest attention to us, they continued their piercing, drawn-out cries, accompanied by a deafening cracking of whips. The Moldavians found particular delight in such rides in those days, but the experience seemed wild to a person unaccustomed to it. In a short time we had crossed the river Jijia (about five versts from Skuliany) above which rises the magnificent castle of the boyar Nicholas Rosetti-Rosnovanu * in all its splendor on the estate Stânca. After racing along for another half hour we

* The father of our close relative, Countess Maria Nikolaevna Sollogub, by marriage.

suddenly stopped. I asked, as best I could, the reason for this stop. "Jumătatea drumului" [half-way] replied the *surugi*. After a brief rest, badly needed by both the horses and the coachmen, the latter taking advantage of it to smoke their little pipes, we raced on again and soon found ourselves on an elevation from which we could see the city of Jassy, beautifully situated on a hilly site, encircled by a winding river, the Bahlui.

The view of Jassy from this vantage point made the most pleasant impression upon me. The morning was clear, the building of the Metropolitan with the church, the picturesque suburbs, flanked by gardens, and particularly the beautiful monasteries, Frumoasa and Cetățuia, close to the city, made an attractive picture. Without a doubt few cities present as beautiful a sight as Jassy. But only at a distance. As soon as you approach the city through the filthy Jewish suburbs all charm vanishes.

However, since I entered the city this time through the gate, Copou, which led directly into a wide street (Main Street) with some of the better homes, I was spared the full impact of this impression. With deafening shouts and a cracking of whips I drove into the courtyard of our consul's home. At a sharp turn in the yard my carriage received such a jolt that the axle cracked in two. That means that I shall remain here for a long time, I thought. And my premonition did not mislead me. I do not quite recall the day of my arrival in Jassy—I think it was the 17th or 18th of October, 1841.

On alighting from the carriage, I immediately asked to be announced to our consul, Karl Evstaf'evich Kotsebu, who received me most cordially and invited me to stay in his home until I could find living quarters. I accepted this offer and began to unpack in the room assigned to me. Here I noticed that the bottom of my strong box was broken and that some Dutch gold pieces which remained from the trip were dropping out of it. I had to pick them up on the stairway and even found some of them in the carriage. This too I took as a bad omen.

However, I did not think about this too long because I was charmed by my friendly reception in the Kotsebu family and by my new friends and co-workers, the secretary of the consulate, Fedor Antonovich Tumanskii,[1] and Alexander Khristianovich

Pekgol'd, our postmaster in Moldavia, where we [Russians] enjoyed the right to receive our own mail, just as we did in Wallachia
and in Turkey. I soon became close to both of them and spent
many years in the friendliest relations with these worthy people.
They were considerably older, but this did not prevent us from
being on an entirely equal footing.

Before dinner I took a short walk through the main streets of the
city and was pleased with the impression they made upon me. I
liked the private homes of the boyars, built for the most part with
a garden on one side and a court on the other. Many bore traces
of eastern architecture—the wide porch and the long roof gave
them a unique appearance. The house occupied by the Russian consulate was of the same style. Only occasionally did I encounter
buildings constructed in the European fashion. Except for Main
Street, the streets were irregular and winding. Next to some of the
large houses the poorest of hovels or mud houses could also be
seen. All this seemed peculiar to me and aroused my curiosity. I
was amazed by the variety of costumes. Many of the older boyars
and even some of the middle class still wore what was then the
national costume—a sort of long Bukhara robe, under which could
be seen a short fur-lined vest, and red *şalvari* [wide trousers]. Some
wore colored Turkish boots and, on their heads, tall caps (*căciuli*)
made of grey Astrakhan and very unusual in shape, resembling a
pumpkin. This mode of attire was, of course, not too conducive to
walking. Consequently, I encountered few of these people on foot.
All of them sat pompously in carriages with a soldier [arnăut] in
the coach box. When employed by the more prosperous owners,
these soldiers, who were armed with daggers, were richly dressed
in coats of red cloth, stitched all over in gold thread. An indispensable accessory of one of these lackeys, no matter what his attire, was
a long chibouk with an amber mouth piece which he always carried in his hands, awaiting the order of his master to hand him the
pipe. In those days the Moldavians were such avid smokers that
frequently they would stop in the middle of a walk in order to
smoke a pipe. I encountered also some dandies, dressed in the latest
Paris fashion. They were people of the new generation who, for
the most part, had received their education in Paris. There were
few pedestrians among them. The young people almost always

drove around in hired Viennese carriages, called abroad, incorrectly as usual, by the Russian word *drozhki*. They are, one must admit, much more handsome and smarter, as well as cleaner than our Petersburg carriages. Almost all coachmen in Jassy, at least of the better type, are Russian fugitives of the Skoptsy sect.[2] These freaks as well as the numerous Jews encountered at each step spoiled for me the otherwise charming picture which delighted me by its novelty.

Listening to the sounds of the people's speech, I was amazed by the large number of Russian, or at least Slavic, words mixed with Latin which comprises the basis of the Moldavian language.

During dinner in Kotsebu's home I was introduced to his family. His wife, a charming and beautiful woman, born Molly Koskul', was a native of Courland and knew no Russian. My conversation with her was therefore in French. She had at that time only two children—August, about six years old, and Ernest, three years old. They were intelligent and alert little boys. Karl Evstaf'evich Kotsebu was at that time no more than thirty-four years old but he seemed even younger. He was, so to speak, a dandy—always dressed in the latest of fashion. He was sociable and liked worldly diversions. Unfortunately, however, he had a passion for cards which later caused him much harm. At that time he was lucky in cards and, as it invariably happens in such situations, he threw his money right and left with little thought about the future. Moreover, he was generous and kind and never refused to help anyone in need.

As a former Lyceum graduate (he was of the 4th graduation year), he received me as a fellow student, and we established the most friendly relationship.

On the day just following my arrival Karl Evstaf'evich introduced me to the hospodar * of Moldavia, Prince Mihail Sturdza, and to the chief dignitaries of the region.

When we came to see the hospodar, he was holding a council of ministers under his chairmanship. However, in spite of this, when the adjutant announced the arrival of the Russian consul to him,

---

* This is the official title of the ruling princes of Moldavia and Wallachia. They were also called *voevodă*, usually abbreviated as *vodă*.

he left the council immediately to receive us. When it is recalled that in European countries not only envoys, but even ambassadors, must sometimes wait for a long time to be received by the minister of foreign affairs, the unparalleled position and influence which our agents enjoyed then in the Danubian Principalities was amazing.

Hospodar Sturdza lived in his former private home, which at that time had not yet been remodeled and little resembled a palace, but the setting was princely: military guard, adjutants, orderlies, pages in handsome uniforms, and a great number of servants. He was about fifty years old then, unattractive in appearance, of less than medium height, with reddish hair, a long face, small eyes, bowlegged, and somewhat awkward in his movements. In addition, he had a husky and unpleasant voice. However, in spite of these physical defects he held himself with dignity and appeared even attractive in conversation. He had a remarkable mind, abilities, and knowledge. But his niggardliness and greed knew no bounds and he was therefore hated by most of his fellow citizens. Nevertheless, the rule of this hospodar was one of the best in the principality, although bribery was extraordinarily widespread under him, and there was no justice in the courts.

Prince Sturdza's first marriage was to a young woman of the Paladi family. He had two sons from this marriage—Beizadele * Dimitrachi and Grigore who at that time were still in school abroad. After divorcing his first wife, he married Smărăgda Vogorides, the daughter of an influential man in the Turkish government, Logofăt Ştefan Vogorides, a shrewd Bulgarian, who also bore the title of Prince of Samos. They say that Sturdza owed his elevation to the hospodarship largely to this marriage.

Princess Smărăgda Sturdza was still then a young and attractive woman. She was serious, even somewhat harsh, by nature, loved family life, and was much attached to her husband. She had

---

* Beizadea, or son of a bei (plural: beizadele), was a title added at that time to the names of the sons of the hospodars. Later they were arbitrarily called princes, passing this title on to their descendants, which was quite incorrect. Thus neither the Sturdzas nor the Ghicas, nor the Suţus, nor the Moruzis have the right to the princely title.

the reputation of irreproachable morals, a great rarity then in Moldo-Wallachian society. The children of this marriage were Stefan, who died in infancy, Mihail, who died in Paris at the age of twenty, and a daughter, Maria, born, if I am not mistaken, in 1848, and now married to Prince Konstantin Alexandrovich Gorchakov, the youngest son of our state chancellor.

During his rule of Moldavia, Prince Sturdza, who already possessed considerable patrimonial estates by inheritance from his father, increased his fortune to an enormous size.

Our visit with the hospodar was long because, according to the custom of the country, we were served *dulceţi* (preserves), coffee, and pipes with magnificent amber mouth pieces. The prince treated me kindly, but, now, of course, I hardly recall my conversation with him.

Of the boyars on whom we later called, some still wore the national dress,* as, for example, *Logofěţi* Alecu Ghica, Costache Paşcanu (Cantacuzino) Conachi. Others, such as Nicholas Sutzo, Nicholas Canta, George Sutzo, Costache Sturdza-Balş, were in European clothes. Everywhere, however, we were served pipes and the same refreshments as at the hospodar's. The subject of our conversation almost everywhere was General Kiselev, whose wise rule of the principalities had left an indelible impression there.

In a few days I succeeded in finding fairly comfortable living quarters on Main Street in the house of a Jewish merchant, Rosenzweig.† The apartment consisted of one large room facing the street and another with windows into the courtyard where Maksim stayed. The furnishing of the apartment did not cost me too much money or trouble. In those days sumptuous furnishings were unknown in Moldavia, and furniture was not as yet imported from Paris. Almost everywhere people were satisfied with Turkish divans and with furniture of domestic production. But then they lived more openly and displayed a hospitality which cannot even be imagined under the present conditions of life.

---

* Prince Sturdza wore European civilian clothing. On special occasions, however, he wore a general's uniform in which he appeared awkward.

† The price for this apartment was forty chervonets or about a hundred and twenty rubles a year.

Thanks to Karl Evstaf'evich I soon became acquainted with the highest society of Jassy, and I began to receive invitations to dinners and receptions from all over. I visited most frequently the elderly widow, Smărăgdiţa Bogdan, who had special days for receptions and for dinners. I always went there with Tumanskii and Pekgol'd, and seldom were there less than fifteen or twenty persons at the table. The old lady, who had spent her youth turbulently in travels abroad, loved to surround herself with young people and gave full freedom to conversation of double meaning. What could not be heard there! On Sundays I usually dined in the home of the *vistier* Nicholas Rosetti-Rosnovanu, a true grandee who lived completely in the European manner. He spent considerable time in Vienna and Petersburg, mingled in the upper circles there, and adopted the manner and way of living of our aristocrats. With his wealth and the fabulously low price of foodstuffs in Moldavia, he had no difficulty in living most lavishly. Nicholas Rosnovanu was a remarkably handsome tall man. He held himself with much dignity, but was at the same time polite and obliging. His European education and worldly tact compensated for the intellectual gifts with which he had not been richly endowed. He was no fool, however, and, although not above reproach as far as his morals were concerned, even his enemies had to give him credit for the honesty and nobility of his character.

In the days of which I speak Rosnovanu was in a false position. In accordance with the almost universal custom prevailing among his compatriots, he divorced his first wife, Catinca Ghica * (the mother of Countess Sollogub) and had an affair with a Wallachian lady, Anica Filipescu, who was married and whom he soon abandoned. Subsequently, he fell desperately in love with the wife of *Logofăt* Costache Sturdza,† Maria, whom he wanted to marry,

* After the divorce she lived in Bessarabia on her large estate "Lipcani," on the banks of the Prut on the very border of Moldavia.

† The boyar's rank was accompanied in the principalities by titles that were perhaps more numerous than in Russia. In Moldavia boyars of the first class were: *logofăt, vistier, vornic, hatman, postelnic,* and *aga.* In official acts the epithet *vel* was added. For example, *vel logofăt* (in French: *Grand Logothète, Grand Vistier*), etc. To the second class belonged: *serdar, clucer,* and others. There was yet a third class, but I do not recall the names

but, since her husband would not give her a divorce, she merely left him and moved in with Rosnovanu. This illegal cohabitation was the cause of even greater scandal because, at the complaints of her lawful husband and by order of the hospodar, a personal enemy of Rosnovanu, the patriarch of Constantinople publicly anathematized the latter and his common-law wife. Thus excommunicated from the Church, the undaunted couple resorted to energetic measures: They went to Austria and sought out a pastor who converted them to the Protestant faith and married them. The most remarkable aspect, however, is that with the assistance of the same hospodar [Sturdza], who made peace with Rosnovanu and quarreled with *Logofăt* Costache Sturdza several years after this incident, the excommunication was revoked and the marriage accepted as legal. The household chapel of Rosnovanu was opened once more and he and his wife were again declared Orthodox! All this, of course, was not done without considerable sums of money being contributed to the supreme guardians of civil and ecclesiastical laws. From this example the state of morals in this region can be judged.

As would be expected, I met mostly bachelors in Rosnovanu's home at that time. However, after his marriage was accepted, all of the society of Jassy took pleasure in gathering at his magnificent receptions.

The circle of my acquaintances became quite extensive. To begin with, in addition to those above, I should mention among the most important persons, the three brothers Paşcan (Costache, Dimitrie, and Mihailaki), who adopted the name of Cantacuzino because their father or grandfather, a Greek Vlasto, bought the [title and name] documents from one of the impoverished Cantacuzinos. They were called Paşcanu after the estate, Paşcani, which also belonged at one time to the Cantacuzinos; the two brothers Mavrocordat (Alecu and Costache), one more Alecu Mavrocordat, Petrache Mavrogheni, Teodor Balş, two or three more Balş, George

---

of the titles. In Wallachia a different order followed. There the first title was *ban*; the *vornic* and *logofăt* corresponded to the Moldavian *postelnic* or *aga*. The title of *spătar*, or army chief, corresponded to the Moldavian *hatman*.

Ghica, Alecu Sturdza (the husband of the queer Elena Sturdza
who called herself now princess, now countess; the sister of Maria
Rosnovanu whose adventures were related above), Nicholas Canta,
Ion and Alecu Cantacuzino, Costache and Alecu Caracaş, Rosetti,
Başotă and Alecu Rosnovanu. All of them were great *logofĕţi*,
great *vornici*, great *vistieri* and comprised the aristocracy of the
region. I shall have occasion later to speak about their families as
well as about the Moldavians generally. But, in the meantime, I
would like to say that I was awed by the beauty of the Moldavian
ladies and by their elegance. I should add also that many of them
were brought up abroad, particularly in Paris, and were well educated.

In Jassy society I also found several compatriots who were left
by General Kiselev to form a Moldavian military militia. Colonel
Sungurov, an intelligent and worthy man, had charge of the Engineering Section. Colonel Mishchenko formed the infantry and
Major Kafendzhi the cavalry. In addition to them, there was also
a native of Serbia, a retired titular councilor, Paul Ivanovich Stoianovich, who had charge of petitions to the hospodar. All of them
married local girls * and settled permanently in Moldavia. They
were glad to see every new Russian, and they received me with real
kindness. I had to dine in the home of each one of them at least
once a week. And since I also had two regular days when I dined
in the home of Kotsebu, as well as many other invitations, I rarely
ate at an inn.

At that time there was only one decent tavern in Jassy. The pro-
prietor was Regensburger, a Jew who made a fortune thanks to the
Russian officers who congregated there regularly during the last
occupation of the principalities by our troops and who spared no
money in drinking bouts. He knew no other topic of conversation
but the Russians whom he had come to like. "*Wann kommen
meine Russen?*" he asked repeatedly. Frequently, in order to amuse
ourselves at his expense, we would assure him that war with Turkey would soon be declared and that our troops were already as-

---

* With the exception of Konstantin Ivanovich Mishchenko who was
married to the niece of Vashchenko, formerly our consul in Moldavia and,
later, in Serbia.

sembling to cross the Prut. The rapture with which the old man would invariably receive such news is hard to imagine. A complete dinner with wine (Moldavian wine, of course, which we liked better than the foreign kind) in Regensburger's tavern used to cost me some five piastres or lei.*

Except when invited elsewhere, I spent practically every evening in Kotsebu's home. Even after the theater, which ordinarily ended around 10 o'clock in the evening, I would drop in to see him and finish the evening there. Many people used to gather in the enormous reception hall of the consul's home, especially on Mondays, when ladies were also invited. There were frequently dances on those occasions—a pleasure to which the Moldavians abandoned themselves with great enthusiasm. On other days the guests played cards. At one end of the reception hall Karl Evstaf'evich would be deeply engrossed in heavy gambling with his companions; I can say that stacks of gold passed from hand to hand. At a respectable distance from them, we would play a most modest game of whist with Kotsebu's wife. Usually our partners were Pekgol'd, Major Grigore Krupenskii, or Major Leon (the hospodar's adjutant), and Mme. Stoianovich.

Among Kotsebu's partners, who also began by playing whist, but invariably ended with faro or *shtos,* were Costache Sturdza (who was then Minister of Interior), Constantin Caracaş, Ion Cantacuzino, who lost the several fortunes which he inherited at different

---

* A piaster or lei in Moldavian is a fictitious unit equivalent to eight kopeks in silver. The principalities had no currency of their own at that time because the vassal states did not have the right to mint money. However, circulating in the country was Turkish, Russian, and particularly Austrian money because the most extensive trade was carried on with Austria. The largest unit was the Austrian *chervonets,* equal to 35 lei. Then followed the Turkish *icosar* or *irmilic,* comprising 14 lei; then our silver ruble, or as it was called there *carboanţe,* equal to 12 lei, then the *sorocoveţ,* or the Austrian *zwanziger,* equaling two and a half lei. Our twenty-kopek piece had the same value. There were also many half *zwanzigers.* The smallest coin was the Turkish *para.* A piaster contained, as far as I can recall, forty of them. In Moldavia I did not come across these *paras,* or *parauţ,* as they were called. But in Wallachia I used to get them often; they always caused me trouble because they were small and thin and frequently slipped through my fingers.

times, and many more of the top-ranking boyars in Moldavia, all of whom loved games of chance. However, of all those present at the gatherings, the one who impressed me more than anyone else by his handsome appearance and aristocratic air, or as the French say, *son grand air*, was Prince Egor Matveevich [George] Cantacuzino, a retired Russian colonel, who lived at that time on his extensive estates in Moldavia.

I should like to pause to consider him briefly. Here, in a few words, are his life history and his personal adventures.

The princes Cantacuzino, as the Comnens and the Paleologis, trace their origin from Byzantine emperors. Ion and his son Matei Cantacuzino, who ruled in Byzantium in the fourteenth century, did not form a dynasty. Nevertheless, their heirs may be proud of the fact that at least two of their forefathers sat upon the throne. For what reason I do not know, but in the sixteenth century one of the Cantacuzinos, the last remaining of his family, left Constantinople and moved to Trebizond on the south shore of the Black Sea. Later two brothers Cantacuzino, no doubt his sons, moved to the Danubian Principalities, one to Moldavia and the other to Wallachia. Both of them were rich and assumed a considerable position in these principalities. To differentiate between the heirs of the Cantacuzinos who settled in Moldavia from those who settled in Wallachia, the former were called Deleni and the latter Măgureni, after the names of their principal estates. Both soon attained such power and importance that they became the greatest magnates or boyars in the land.

Both in Wallachia and Moldavia some members of their family were even hospodars. In consideration of the great services rendered by them in the struggle with the Turks, the Cantacuzinos and the Besarab-Brâncoveanus, one of whom also ruled in Wallachia, were the only families to have their princely title recognized by Russia and Austria. After the unfortunate campaign of Peter the Great on the Prut both Cantacuzinos who participated in it, as well as Cantemir,[3] had to move to Russia.* In a few years the Cantacu-

---

* Prince Brâncoveanu, then the hospodar in Wallachia, who pledged support to Peter the Great, betrayed him at the last moment, a fact that was

zinos returned to the principalities. However, during the reign of
Catherine the Great, who was incessantly at war with the Turks,
one of his descendants, Prince Matei Cantacuzino (Deleni), was
also obliged to seek asylum in Russia. Since all his estates in
Moldavia had been confiscated by the Turks, Empress Catherine
presented him with an estate of one thousand serfs in Vitebsk
province, where he took up residence. He was married to the
daughter of a Moldavian hospodar, Prince Callimachi, and he had
three sons by her, Alexander, Grigore,* and George—the Egor
Matveevich referred to here. During his residence in Russia Matei
suffered a stroke of paralysis which left him in a pitiful state. All
parts of his body were paralyzed. His mental faculties left him,
and he lived for almost twenty years in this state of cretinism. In the
meantime, as a result of the peace concluded with Turkey, the
sequestration of his estates was lifted. Because of the minority of
his children, his property in Moldavia was put in trust under the
closest relatives of the unfortunate Cantacuzino, his wife's relatives, the boyars Sturdza and Ghica. Both men were married to
the sisters of Cantacuzino's wife, born Callimachi. To obtain an
idea of the immensity of the property that was placed in trustee-
ship, suffice it to say that Cantacuzino could travel from his village, situated near Khotin, on the shores of the Dnestr, to Galatz
itself, near the estuary of the Danube, without leaving his lands.
This distance may be estimated at more than 300 versts. Moreover,
he owned estates in the Carpathians (in Upper Moldavia) and in
Bukovina. The trustees managed affairs in such a way that half of
the estates were put up for sale at public auction, and they themselves bought the lands for a song. They succeeded in this outright

---

largely responsible for the unsuccessful outcome of the campaign undertaken by him against the Turks.

* Prince Alexander Matveevich married Elisaveta Mikhailovna Daragan
and had a large family about whom I shall have occasion to speak later. He
lived alternately in Russia, Moldavia, Greece, and Dresden, where his children were educated. He died in Moldavia. Prince Grigore Matveevich served
as colonel in our Guard and died a heroic death in the Battle of Borodino.
He was a man of great promise who was distinguished by his fine mind,
wide knowledge, and also his nobility of character. He was a bachelor.

plundering by bribing the Moldavian officials who helped them. The majority of the estates were in Bessarabia, which did not as yet belong to Russia.* Drawn into the deception of the trustees, the wife of Matei Cantacuzino did not protest against this sale, and her sons were still children.

On becoming of age, the young princes Cantacuzino started legal proceedings, but since the trial had to be conducted in Moldavia, where the chief boyars were biased in favor of the thiefs, there was practically no hope of winning the case. Nevertheless, the trial took place and the plaintiffs were subpoenaed to appear in court. Prince Egor Matveevich was serving at that time as cornet in the Horse Guard Regiment. He rushed to Jassy, made his appearance in court, and, realizing that the case would be lost, showered the worst curses upon the judges and the Moldavian government. After this scandal, he hurried back to Russia. With the annexation of Bessarabia to Russia, the princes Cantacuzino attempted to obtain justice through the Russian government. However, because of a special treaty concluded with Turkey, we were not able to reopen cases already acted upon by the Moldavian government which involved payments of money in that area. Thus the Cantacuzinos lost a good half of their former fortune.

After serving for some time in the Horse Guard Regiment, Prince Egor Matveevich transferred to the Engineering Corps which was more in keeping with his education and gifts. He attracted the attention of the well-known General Benigsen who chose him as his adjutant. He remained with him through the campaigns of that great epoch and earned all marks of distinction which he could possibly obtain according to his title and rank. At the conclusion of the campaign, he returned from France with the rank of colonel,

---

* Part of these estates belonged to Countess Maria Nikolaevna Sollogub who inherited them from her mother, born a Ghica, the daughter of one of the trustees, as well as to her sister, Baroness Rozen, who owned the magnificent estate Lipcani, the former property of the Cantacuzinos. However, most of the plundered Cantacuzino estates were in the hands of the former Moldavian hospodar, Mihail Sturdza, the son of the other trustee. I have had occasion to pass through these estates and, knowing the manner by which they fell into the hands of the present owners, I cannot help but grieve for the unfortunate Cantacuzinos.

and while still young in years was given the command of the Second Bug Uhlan Regiment. As we can see, his career was a brilliant one and would have taken him far if special circumstances had not cut it short in a most unusual manner.

In 1821 the Greeks rebelled to throw off the Turkish yoke. To prepare a useful diversion for the main revolt, the so-called Heteria [Philike] was formed, composed of the Greeks and Arnăuts residing in the Danubian Principalities. Thus two great detachments were formed—one for action in Wallachia and the other for Moldavia. Elected as commander of the former was Alexander Ipsilanti, a native Greek, Russian trained, and of the latter, Prince Egor Matveevich Cantacuzino. Both of them launched upon this desperate venture without asking the Russian government for permission. In the fervor of his Greek patriotism, Prince Cantacuzino rushed to the battlefield without even turning over his regiment to his superior officer in the proper manner. Like Ipsilanti, he was confident that Alexander I, who sympathized with his coreligionists, would not condemn his action.

Everyone knows the unfortunate fate of the Heteria. The Turks directed tremendous forces into the principalities and crushed it, first in Wallachia and then in Moldavia. Engaged in active combat practically every day, Prince Cantacuzino retreated to the Prut with a handful of men and stopped overnight in the castle Stânca, which I saw on my arrival in Moldavia. The Turks followed on his heels and camped near Stânca. By morning preparations were completed for the final desperate, but, as far as the Greeks were concerned, completely futile battle. In the meantime, across the way, in Skuliany, on the very frontier, were our troops and with them were Generals Inzov and Kiselev, watching the Turks. Cantacuzino's wife arrived there. Anxious about her husband and realizing that he faced an inevitable death, she pleaded with Kiselev to order Cantacuzino to come to Skuliany for negotiations. No sooner had he arrived there than the Turks furiously attacked the unfortunate Greeks. Only a few of them had time to save themselves and swim across the Prut to our frontier. An eye-witness of this scene, Alexander Grigor'evich Rozhalin, whom I saw so often in Skuliany, related all of its horrors to me. The Turks mercilessly killed even the utterly exhausted men and the entire river was red with blood.

As was to be expected, the Greeks accused Cantacuzino of treason. But if he crossed over to Skuliany, he did so with the hope of finding some assistance for them. The rapid finish of the battle prevented him from realizing this hope. Moreover, our authorities held him in Skuliany. Needless to say, this was the most agonizing moment of his life.

The news of the Heteria and its tragic fate reached Emperor Alexander in Laibach during the congress where, under the influence of Metternich, the emperors were discussing the question of the restoration of peace and order in Europe which had been upset by various revolutionary organizations. At such a moment and before the others at the congress, he naturally had to demonstrate his displeasure and to subject to punishment the Russian officers who had participated in the revolt against Turkey. The Austrians themselves dealt with Prince Ipsilanti. Since he had the misfortune to seek asylum on Austrian land, he was immediately imprisoned in the Spillberg Fortress where he died shortly afterward. Prince Cantacuzino, however, was on Russian soil. In spite of the demands by the Turks and Austrians, the emperor limited his punishment to an order restricting the prince to residence on his estates in Bessarabia where he lived for about ten years. Later he was permitted to travel to Odessa and finally was granted freedom to live anywhere he chose.

After this unfortunate campaign, Prince Egor Matveevich made his residence with his family in one of his villages, Markautsy, in the district of Khotin. If I am not mistaken, he was married in 1811 to Princess Elena Mikhailovna Gorchakov.* Their contemporaries who knew them personally told me that the young couple dazzled everybody by their arresting beauty. However, this marriage could not be considered a happy one. The virtuous, modest, and

---

* The sister of Prince Alexander Mikhailovich, now state chancellor. He had three more sisters, one of them married to Count Sollogub, the other to Khvoshchinskii, and the third to Obol'aninov—all of them said to be beautiful. I know little about their father, Prince Mikhail Gorchakov, except that he was married to Baron Osten-Saken's widow who was born Countess Ferzen, and that because of this lived for a long time in the Baltic province. There Prince Alexander Mikhailovich received early education before entering the Lyceum in Tsarskoe Selo in 1811.

gentle Princess Elena Mikhailovna, this angel of kindness, as everyone justly called her, suffered much from the whims of her husband. There was much in his character that was noble and even chivalrous, but he was flighty, willful, and arrogant. His extravagance at times knew no bounds. The command of the Bug Uhlan Regiment cost him tremendous sums of money. Once in an effort to produce a brilliant display by his regiment during the emperor's inspection, he bought horses for all of his soldiers that even officers would be proud to parade. On another occasion during a march with his regiment he happened to stop for a rest in a place where there was no firewood to cook *kasha* for his soldiers. Without hesitating a moment he ordered his own costly carriages to be chopped up and used to kindle a fire. I heard many similar anecdotes about him. I am consequently not surprised that he was unable to preserve his great fortune. Moreover, he was a passionate gambler and always lost. He had no opportunity to squander money in Markautsy, but frequently he would set out from there with sacks full of gold and go to Odessa where he played for high stakes and always returned with his pockets empty.

When full freedom of movement was granted to him, he moved to Moldavia where he still owned the magnificent estates Hangu, Bălţăteşti, and Largu. He sold the latter shortly after his arrival to pay his card debts, which he always honored promptly. He sold the home he inherited in Jassy for the same reason. It was acquired for a song by Stoianovich who managed his affairs for a time.

This time Princess Elena Mihailovna did not remain long in Moldavia. She wanted to return to quiet Bessarabia, and she settled with her younger children in the village Ottaki which belonged to the older brother of her husband, Prince Alexander Cantacuzino, and where his much-respected widow, Princess Elisaveta Mikhailovna, lived with some of her children. Princess Cantacuzino had a sincere affection for her. Both families were congenial and spent many years together there in peace and quiet. Fate has brought me so close to the Cantacuzinos that I shall have occasion to speak about them frequently in these notes. For the present I shall speak about Prince Egor Matveevich, one of the first of the Cantacuzinos, as I said above, with whom I became acquainted on my arrival in Moldavia.

He was at that time about fifty-seven years old. In spite of the tempestuous life he had led, he still appeared youthful and took part in the gaieties and even pranks of the young people who surrounded him. His uninhibited speech (*franc parler*) and his unceremonious treatment of everyone often confused those present, especially the ladies, whom he delighted in embarrassing. Whatever was in his mind was on his tongue. However, since he was witty, his pranks were funny and original. He was unusually short-tempered. Therefore, it was always dangerous to argue with him. The Moldavian boyars and even the hospodar himself feared him greatly. Once, when playing cards with Prince Sturdza, who was his first cousin, he had an argument with him and showered such abuses upon him, reminding him of all of his intrigues and dishonest transactions, that the boyars present, who always treated the hospodar with servility, were so afraid and embarrassed that they did not know what to do. However, all this always ended happily for Cantacuzino. As for me, I stood in the good graces of Prince Cantacuzino. He was never angry with me, even when I was obliged to contradict him. In his good moments he was extremely amiable and was liked for his keen and fine mind and his noble manners and courtesy.

Thanks to the wide circle of acquaintances that I acquired shortly after my arrival, I began to lead the life of a man of the world, a life much more diversified than I had led in Petersburg. This, however, did not prevent me from spending several hours a day in serious study. In the morning I went to the offices of the consulate where I studied the official correspondence and the archives. After dinner, before going out anywhere for the evening, I read political books. I also made a particularly thorough study of the French language. I spent several years pouring over the *Grammaire de Grammaires*. In order to perfect myself in diplomatic style, I took notes and made résumés from political articles in French journals.

There was quite a bit of work in our consulate in Jassy then. Although the Danubian Principalities were under the supreme sovereignty of Turkey, the vassal dependence upon the Porte was so weak that it was reduced merely to the paying of tribute. The Turks did not dare to interfere in questions of internal administration in Moldavia and Wallachia. The principalities are indebted for this favorable position to Russia alone who shed much blood to free them from the Turkish yoke. At the close of each war we gained some new concession in our treaties with the Porte. Finally, by the Treaty of Adrianople, concluded in 1829, the Turks were completely banished from the principalities. They were even forbidden to take up residence there, and their fortresses on Moldo-Wallachian territory, Giurgiu and Brăila, were leveled to the ground. This was the greatest blessing for the principalities because the Turkish troops had made forages on the villages from these fortresses and had taken foodstuffs and cattle, which they had occasionally paid for, but at a most trivial price. As a consequence, the magnificent estates along the Danube, which are now the granary of Europe and a source of wealth to the principalities, yielded no profit at that time. Moreover, because of the Treaty of Adrianople, the hospodars of Moldavia and Wallachia, who heretofore had been appointed by the Porte, a condition which gave rise to terrible abuses,* now had to be elected by the inhabitants of the principalities from their own compatriots. At first, however, it was decided to make an exception to the rule in one respect, namely, that there should be no elections because it was assumed that the inhabitants were not sufficiently prepared for it. Because of lack of experience and other reasons, they might make a poor choice. Consequently, with the introduction of the new order Turkey and Russia, by mutual agreement, appointed as hospodar of Moldavia the *vistier* Mihail Sturdza and in Wallachia the *spătar* (army chief), Alexander Ghica. We shall see later to what extent these choices proved to be fortunate. But, in the meantime, suffice it to say that a new administration in the principalities based on the idea of self-government followed in 1831. It was proposed by our government and was submitted for discussion to

---

* The Porte usually appointed as hospodars faithful Greek Phanariots, and of those only the Moruzi left a good reputation. Others, such as Caragea, Suțu, and Mavrocordat, thought more of their own advantage and deluged the region with their relatives and friends, all of whom succeeded in growing rich. The Greeks were soon hated in both principalities. Under the Phanariot hospodars Greek became the language of the upper classes. Later it was replaced by French. I found that many boyars still frequently spoke Greek among themselves.

the divans, or chambers of Moldavia and Wallachia, which were comprised of delegates of the region under the chairmanship of General Kiselev. In this constitution, or Organic Statute, as it was called, all regulations dealing with the administration of the principalities were worked out to the minutest detail. This Statute, which made General Kiselev famous, was approved by both chambers. A gift from Russia, it was received under her protectorate or guaranty, and Russia became officially the *Protectorate Power of the Danubian Principalities*. Our consuls were instructed to make sure that the regulations were strictly adhered to. This political aspect of their duties was of great importance and gave them an exceptional position in the principalities.

Furthermore, consular jurisdiction, which exists to this day in Moldavia as well as everywhere in Wallachia, and which gave the consuls the right of trial and punishment over their fellow citizens, caused us many difficulties in Jassy.

Nevertheless, our greatest burden of work was the correspondence with the authorities of our border and of Bessarabia. The annexation of Bessarabia made the landowners dual citizens (*propriétaires mixtes*), that is, persons who owned estates in both Moldavia and Bessarabia. Each one of them had some sort of a court case—life is not worth living to a Moldavian unless he has a court case. All this was referred to the consulate, and we had to be the intermediaries in the foulest lawsuits before tribunals (I am referring to Moldavia as well as, unfortunately, to our Bessarabia) where one could not find a single honest judge.

It can well be concluded from this that our diplomats in Jassy passed through a school unlike any other in a European embassy. It is strange that so few of them rose in the world—probably because not everyone had the necessary luck and also because of office squabbles which diverted them from political questions. I tried to avoid this danger and always followed political events carefully.

My kind and clever friend Tumanskii was terribly lazy and was always grateful when I took over his work. For my part, after the dull service in the Asiatic Department, where I performed practically the duties of a scribe, I welcomed the pleasant test of my strength in a more important field. Moreover, I never abandoned

St. Petersburg, Nevskii Prospekt, in the 1840's

Moldavian Post in the Early Nineteenth Century

hope of following a diplomatic career and prepared myself assiduously for it, in spite of all the failures.

In addition to the regularly appointed officials in our consulate,
we also had many civilians drawn from local and Bessarabian resi-
dents who knew the Russian and Moldavian languages. They performed the larger part of our work. I recall the kind and honest
Harito, the conscientious and niggardly Kirisha Trofimovich Komarnitskii, and the gifted and nimble Evdokim Andreevich Ianov
who was my first teacher in administrative and legal matters. Had
he received a decent education, he would have made an excellent
statesman. There were others also, such as the older Makedonskii,
Tomida, and others, but I had less to do with them. Harito, for
whom I developed a deep affection, died about two years after
my arrival. He was of Greek extraction and a great admirer of
Capodistrias * about whom he always spoke with enthusiasm. Well
versed in classical Greek literature, the enthusiastic Harito liked to
quote passages from the tragedies of Sophocles, and he did it with
a passion which at times reached madness. He loved particularly
the tragedy of Oedipus, which he also knew in the Russian translation of Ozerov. How often the kindly old man, assuming a tragic
pose, which made us laugh, would surprise us by exclaiming:

> You see my head that has lost its hair
> Sorrow has dried them—and the wind has blown them away.

This honest man, who had so many chances to get rich by taking part in lawsuits in a country where bribery was so widespread,
lived and died in poverty. Fortunately, he had no family.

About two months after my arrival, our consular staff was increased by two more official employees who arrived from Petersburg. They were Shchulepnikov and Sokolov, who had just gradu-

---

* Count Capodistrias, one of the most remarkable people of his time,
followed Prince Czartoryski as minister of foreign affairs during the reign
of Alexander I. Participating actively in the fate of his fatherland, Greece,
he contributed much toward the recognition of her independence and was
selected head of the Greek government with the title of president. Despite
the great service rendered by him to Greece, he fell by the hand of a fellow
countryman. The assassin, Mavromichalis, belonged to one of the most
distinguished families of the region, who looked with envy upon his rise.

ated from the Oriental Faculty, where they had studied the
Moldavian language under a friend of mine, Iakov Danilovich
Ginkulov. The department teaching this language in Petersburg
University was eliminated shortly after that. Evgraf Romanovich
Shchulepnikov, no fool but a highly flighty young man, proved
to be a gay companion for us. However, because of the types of
recreation he chose, he mingled with people for whose company
I did not care. His friend, Kapiton Aleksandrovich Sokolov, a
lanky skinny youth, awkward and with a stutter, did not impress
me favorably. But after I had an opportunity to appreciate his
excellent qualities, he became my good, I can say, my bosom friend,
and our friendship continued until his death in 1859.

I liked the life in Jassy, particularly after the evening winter entertainments began. The first large ball was given by the hospodar
on December 6th on the occasion of the name-day of Emperor
Nicholas Pavlovich, the most august patron of Moldavia. That day
was always celebrated with unusual festivity. After the mass and
services in the Church of St. Spiridon (near the consul's house)
where the metropolitan himself with all the bishops conducted the
service in the presence of the hospodar, all the officials, and a great
number of boyars, they all proceeded to the consul's house to con-
gratulate him. A *postelnic*, or secretary of state [in charge of foreign
affairs], representing the hospodar, arrived in a princely gala carriage, surrounded by a platoon of uhlans. A military band played
during the reception in the courtyard of the consul's home. In the
evening the city was illuminated, and the celebration was concluded with a gala ball in the home of the hospodar.

All of Jassy's upper and middle-class society was present at this
ball. I was lost in admiration at the abundance of diamonds worn
by the ladies, particularly the middle-aged ones. Many of them had
diadems which at the present time could not be purchased for less
than thirty or forty thousand rubles, and some displayed even more
expensive ones. All this was inherited wealth. As everywhere in
the East, the passion for precious stones was at one time widely
prevalent in the principalities. The clothing worn by the ladies was
elegant and made in the latest fashion. The Moldavian ladies are
known for their vivacity, good minds, and ease and pleasantness
of manner, but few of them have naturally noble manners, that

innate distinction which is hard to acquire from others. The women as I have already said are very beautiful. One encounters few or practically no homely faces as in other European gatherings, not excluding our Petersburg. As for beauties there were a great many. I should like to mention among them Mme. Callimachi, born Paşcanu,* Mme. Casimir, born Kogălniceanu, and the two daughters Spiro. All of them were brunettes. There are practically no blondes in Moldavia. However, Mme. Callimachi was an exception. She and her sister were the best educated and most well-mannered ladies in Moldavia. The same could be said about the men also. I was amazed at their ability to dress in good taste. I speak of course about the young men who were educated in Paris. The middle-aged men who had discarded their national attire only recently were not at ease in European clothing. This change in attire occurred no more than about ten years before my arrival, under the administration of General Kiselev—that true reformer of the principalities. It is particularly remarkable how the young men adopted the military full-dress coats which they wore with genuine elegance. Among the young men my attention was attracted particularly to Ion Ghica (to differentiate him from the other numerous Ghicas, he was called *brigadier*, because his father had the Russian rank of state councilor, corresponding to the military rank of brigadier, a rank abolished long ago); Alexander Sturdza, nicknamed Koko, the first dandy in both principalities; Petrache Mavrogheni, the most decent and sober of all Moldavian young men; Vasile Alecsandri, a remarkable poet; his brother Ion Alecsandri; Major Leon (the last two were adjutants of the hospodar); Major Grigore Krupenskii, with whom I soon struck up a great friendship; Vasile Rosetti, who succeeded in squandering a tremendous fortune in Paris and Vienna, and the two Docan brothers. I spent many years in the company of these young people and my relations with them were always pleasant.

Shortly after the celebration Kotsebu began his preparations to go to Bucharest for an interview with the consul general, Dashkov. He offered to have me accompany him, and I gladly accepted.

---

* This is the older sister of Pulcheria Arghiropulo, our good friend. They are daughters of the boyar Costache Paşcanu, wrongly called Cantacuzino.

The weather was not favorable for this trip. Winter had not as yet set in, but the dampness had made the ground so wet that in some streets in Jassy, which were cleaned poorly or not at all, the mud was up to one's knees. Out in the field the road was abominable. In spite of this Kotsebu decided to go because he had scheduled his time in such a way that he would return for Christmas. We started out before dawn. Our two-seated carriage was harnessed with twelve horses, guided by three *surugi*. Since the horses were hitched in pairs, one after the other, the first pair was far out in the street when the carriage still stood in the courtyard in front of the doorway. We started out accompanied by horrifying howls from the many mounted *slujitori* [guides] escorting us. In those days a journey by the Russian consul was conducted in a most elaborate and spectacular manner. A courier rode ahead of us to prepare the horses. In the cities we were met by the entire detachment of the gendarme command and sometimes by the *ispravnic* (prefect as it is now called), where either overnight accommodations, or dinner, or some sort of refreshment were always kept ready. We had no occasion to stop at taverns. Besides there were no decent ones at that time in the principalities. I was amazed by the generosity with which Kotsebu showered money upon the *surugi*, the *slujitori*, and the others who waited on us. He thought that it was necessary to do so to maintain the consular dignity. However, the costs for such trips made in line of duty were entered as extraordinary expenses of the consulate and were paid by the treasury. In spite of the great number of horses, we barely dragged along through the swampy mud, and after brief stops in the district towns Vaslui, Bârlad, and Tecuci, which seemed to me quite poor and dirty, we reached Focşani only on the following day toward evening, having crossed the river Siret. The river Milcov which forms the frontier between Moldavia and Wallachia flows almost in the center of this city.

Except for some parts between Jassy and Vaslui, and perhaps near Bârlad, where fairly attractive houses can be seen as well as small groves and orchards, all the rest of the area to Focşani is a steppe which is boring to the traveler. Only here and there, near the towns, did we encounter some trees. There is no sign of vegetation in the villages and small towns at this time of year. This entire countryside resembles the south of Russia. Just as there, a well with an enormous lever and a pail to draw water is at the entrance of each small town.

In Focşani, situated almost half-way between Jassy and Bucharest, we were met by the Wallachian authorities. In spite of the close kinship between the neighboring Rumanian nations, some difference is apparent in their appearance, clothing, and speech. A Wallachian impressed me as being softer and more affable than a Moldavian.

We entered the territory marked by the victories of Suvorov [4] over the Turks. His first victory was near Focşani, and the second and decisive one over the grand vizir, near Râmnic, for which Catherine the Great bestowed upon him the title of Count of Rymnik (Râmnic).

Râmnic is a Wallachian district town, one station from Focşani, near a river of the same name. This mountain stream, which a chicken could cross in the driest season of the year, inundates the area over a great distance during periods of flood, and then the crossing of the river is sometimes accompanied by great danger. Suvorov's only son, Lieutenant-General Prince Arkadii Alexandrovich, was drowned in this way in the Râmnic. He commanded a cavalry division during the war with Turkey which ended in the Peace of Bucharest. This occurred in April, 1811, that is, about twenty years after the victory gained on the same spot and for which his father received the title of Count Rymnik. Near the river stands a watchtower in commemoration of this unfortunate event.

The road continues through the steppe, which stretches all the way to Bucharest. It is even more dreary and vast than the Moldavian steppe. For about three or four stations after Râmnic we dragged through the same awful mud until we reached the city of Buzău, noted for the magnificent building which belongs to the episcopal see of the place. Travelers usually stop there, and a most hospitable, even sumptuous, reception awaits them. The building was erected for this purpose by order of the government and is paid for by church funds, which are extremely large in the principalities. We did not find the bishop himself in Buzău. He usually spent the winter months in Bucharest where he participated in the sessions of the legislative assembly, of which he was regarded as an indispensable member. An eparchial bishop of Buzău in the ecclesiastical hierarchy of Wallachia occupies the first place after the metropolitan. Here too, close by the city, a river by the same name flows which floods almost more than the Râmnic. It is absolutely impossible to build a bridge across these torrents, and one must wade across them. At the time, the water had risen considerably and we crossed the river with difficulty. The water reached above the wheels.

From Buzău, where we were received with great esteem and where they gave us an excellent meal, we left at night in such pitch darkness that it was impossible to distinguish any object within two steps. We lost our way several times, although we were accompanied by *dorobanți* (the gendarmes in Wallachia who correspond to the *slujitori* of Moldavia). Our carriage swayed from side to side and finally overturned in a ditch. Kotsebu was the first to be thrown out, and I fell on him. By some strange chance neither of us was hurt. Our bruises were minor and, once up, we continued on our way as if nothing had happened. In the morning we reached the nearest small town, Urziceni, on the shore of the Jalomița River, across which our engineers built a fairly decent bridge during the last occupation by our troops. The distance to the Urziceni station is the longest I had ever occasion to see in the principalities. It could easily take the place of two. From there, if I am not mistaken, only three stations remain to Bucharest, where we arrived late in the evening.

We drove in through the Colentina gate, named after the estate situated nearby with a magnificent manor house, Colentina, belonging then to the *beizadea* Grigore Ghica. A wide and long street, Pogul Targul-de afară, runs from the gate. We drove along it for about twenty minutes until we turned off into another street. I could judge from this that Bucharest is considerably larger than Jassy. Even then it had more than a hundred thousand inhabitants, whereas the Moldavian capital had only approximately fifty thousand.

Kotsebu dropped me off at the hotel "Casino" and he proceeded on to the Russian consulate general where an apartment was prepared for him.

I was assigned a fairly decent room at the "Casino." To reach it

I had to go through a large hall with a gallery. It served for the public balls of the middle class, and the desk clerk who received me told me that such a ball had taken place there the night before our arrival. Fortunately, during my brief stay in Bucharest there was not a single large function in the "Casino," and I could enjoy peace and quiet in my room. However, it served me merely as a place to sleep, because during the day I was occupied paying visits.

On the following morning I sent for a carriage and went to pay a visit to our consul general, Iakov Andreevich Dashkov. He made the pleasantest impression upon me. I liked his noble manners and restrained politeness. He could pass more easily for an Englishman than for a Russian. This is not surprising because he was born in America when his father was ambassador to the United States and had received an entirely English education. Several years spent in the Pages Corps could not efface his English mannerisms. He spoke Russian rather badly. Having no inclination for military service, Iakov Andreevich entered the diplomatic field. He began, I think, with an appointment to take charge of the foreign correspondence of the commander in chief of our squadron in the Archipelago, Count Geiden. He was attached to Geiden during the famous battle of Navarino. Thereafter Dashkov was secretary to the mission in Copenhagen for a long time. Then he was appointed counselor to the mission in Constantinople. However, he never served in that capacity because no sooner was he appointed to this post, than he was asked to accept the administration of the consulate general in Bucharest because of the departure of Vladimir Pavlovich Titov, who had to manage the mission in Constantinople in the capacity of chargé d'affaires in the absence of Ambassador Butenev. Titov never returned to the principalities because he was appointed shortly thereafter ambassador to the Ottoman Porte. And Dashkov continued in his position of consul general for almost seven years.

Iakov Andreevich received me in a kindly manner and invited me to dinner. In the meantime, I went on a tour of the city. In Bucharest, just as in Jassy, almost all cabmen are Russian Skoptsy. I found it a great help to be able to talk to them. We set out directly for the Mitropolie, situated on a considerable elevation from where one can see the entire city. Bucharest appears from there to be very picturesque and reminds many of Moscow. Except

for the main street, Podul-Mogoşoaiei, which runs in a straight line
through the center of the city, also Podul-Targul-de-afară, already
mentioned above, practically all streets * run in uneven lines, crossing in many places the winding river, Dambovița. Most homes are
built between a garden and a courtyard. Here and there one encounters shanties, but they are incomparably more presentable
and neater than those in Jassy, thanks to the absence of Jews who
at that time were almost altogether forbidden to take up residence
throughout Wallachia. This element is replaced here by the Bulgarians, Serbians, Hungarians, Greeks, Armenians, and Albanians.
Just as the Jews in Moldavia, they took into their hands practically
all industry, commerce, and trades. Aside from their other advantages over the Jews, these Christian nationalities are particularly
distinguished from them by neatness, beautiful costumes, and
noble appearance. This is noticeable in the Serbians, more than in
any others, and also in the Bulgarians. The Montenegrins surpass
them even more in this respect, but their number, of course, is
insignificant. There the Hungarians engage for the most part in
cattle-raising.† The army in Wallachia wore the same uniform as
that in Moldavia, the only difference being that the collars and
cuffs were yellow instead of red. The cavalry, composed of uhlans,
wore, however, cuffs of white cloth. Our Russian military men
used to say that the uniform of these cavalry men was exactly the
same as that of our Kharkov Uhlan Regiment. In general, traffic
in the city was considerable. There were a large number of carriages, because, with the absence of sidewalks and because of the
impassable mud, only the rabble ventured on foot. The variety of
costumes in Bucharest was even greater than in Jassy.

The house of our consul general is on Podul-Mogoşoaiei in the
heart of the city. It formerly belonged to a prosperous Serbian

* The word *ulița* is used only in Moldavia (it is now replaced there by the
word *strada*). In Wallachia *pod* is the word used for street. Actually it means
bridge.

† There was a small Jewish colony in Bucharest. It was composed chiefly
of the so-called Spanish Jews, who are wide-spread in the East and who in
all respects stand incomparably higher than the Polish Jews. There are very
respectable people among them, bankers for the most part. As for the gypsies,
I will have occasion to speak of them later.

prince, the famous Miloš Obrenović,* who during his banishment lived in Wallachia where he owned great estates. Dashkov bought this home for only ten thousand chervonets. It is now worth twice if not three times as much.

A plain thick stone wall separated the enormous courtyard from the street. Low out-buildings were around the courtyard; they consisted of barns, stables, a kitchen, servants' quarters, and an office. The house itself, a one-story structure of antiquated architecture, was distinguished by its high roof. Behind the house was a fairly pretty garden, descending in three terraces, so that from this side the house gave the appearance of standing on a cliff, and the view from the balcony out over the low sections (mahala) of Bucharest was extremely picturesque.

During dinner I became acquainted with Dashkov's wife who was then still a young woman. She appeared to me to be shy and somewhat frightened of her mother-in-law, Dashkov's mother, a woman of about fifty-five, who obviously ruled the household. Her devotion for her son from whom she never parted was, of course, a fine trait in her character. However, unfortunately, this feeling was expressed by a kind of despotism, by domineering over everything that concerned him. Whether from force of habit, or because of weakness of character, or of love for her, Iakov Andreevich himself never felt this a burden and gave his mother full freedom to rule over his household. To be sure, Evgenia Osipovna was an intelligent, well-educated, and energetic woman, and her son could depend upon her to take over the management of his home.

Iakov Andreevich himself had no great means, I think, but his wife, Pavla Ivanovna, born Begichev, brought him a fairly large dowry which, because of his thrift, he augmented considerably. At that time (December, 1841) they had only one daughter who was, if I recall correctly, not more than a year old.

There were not too many at the dinner table, but in the evening the upper class of Bucharest society gathered in Dashkov's home. There I was introduced to all: Filipescu, Florescu, Văcărescu, Herescu, Bibescu, Băleanu, Bălăceanu, Otetelişanu, and others. In

* Miloš Obrenović was prince of Serbia from 1815 to 1839, when he was ousted by Alexander Karadjordjević. Miloš was restored in 1858.

Wallachia as well as in Moldavia, these endings in *escu* and *anu* are characteristic of names of purely local Rumanian derivation. Native names ending in *un* and *eu* are also encountered. All other names, or practically all, are of foreign derivation. Thus Cantacuzino, Moruzi, Sutzo, Mavrocordat, Mavrogheni, Caragea, and others, are of Greek origin. Sturdza and Balş are of Hungarian derivation; Ghica of Albanian or Arnăut; Catargiu and Ştirbei are no doubt also of the same origin; Rosetti is apparently of Italian origin; Krupenskii of Polish. I am naming only the most important boyars' names in both principalities. Nobility, as a matter of fact, is nonexistent there, and they have no heraldry of any kind. Only persons having some rank are called boyars. In Moldavia all those from *logofăt* to *aga*, and in Wallachia from *ban* to *clucer* (if I am not mistaken) belong to the *boieri mari*. All other titled persons were called *boierinaşi* or, using a contemptuous word, "ciocoi." To be sure, the most honorable titles and ranks were the possession of but a few rich families who formed the aristocracy, but it enjoyed no special hereditary rights. If persons bearing the names of Sutzo, Ştirbei, Ghica, Sturdza, and some Cantacuzinos assume the princely title abroad just because one of their ancestors was a hospodar, this is entirely wrong and they are successful in their pretentions only because of the ignorance of foreigners. As I have already noted, the princely title of *beizadea* is given only to sons of hospodars, as a courtesy (*par courtoisie*). This title, however, can not be passed on to the third generation. I explained above the conditions under which only one branch of the Cantacuzinos and the Besarab-Brâncoveanus have the right to the princely title. In Wallachia both of these names have, as a matter of fact, vanished. Those who still bear this name there received it through the female line. The present Brâncoveanu, the son of George Bibescu (formerly a hospodar for a time, as we shall see) was adopted by the last Prince Besarab-Brâncoveanu and inherited from him an enormous fortune. I later met him and became friendly with him. At the time about which I am writing, he was still being educated in Paris, and his father was in Bucharest. I met him in the home of Dashkov, where I also met Barbu Ştirbei, the brother of George Bibescu. Both of them played a prominent role in their region and achieved European fame. It is, therefore, well to say a few words about them.

Their father belonged to the rank of secondary boyars and lived in Craiova in Little Wallachia, where he engaged in various transactions and amassed a considerable fortune. Bibescu, an intelligent man, recognized the advantages of education. Acting in advance of his still crude compatriots in this respect, he sent his sons, Barbu, George, and Ion, to be educated in Paris. The two older brothers were gifted young men. When they had completed their studies, they returned to their fatherland at the time when our government, after delivering the principalities from the Turkish yoke, planned to give them an independent government based on European principles. General Kiselev was quick to notice these young men, who were so strikingly outstanding compared with other young men of their age. He employed them in his service, and they soon became his chief assistants in launching the reforms. Because of the position which they thus acquired, both of them made excellent marriages. The elder, Barbu, married the daughter of the Moldavian boyar, Paşcanu (Cantacuzino). In addition, he was adopted by a childless rich man, Ştirbei, who willed his estates and all his fortune to him. The second, George, married the daughter of a Moldavian boyar, Alecu Mavrocordat, and the niece, if I am not mistaken, of the last of the princes Brâncoveanu. The latter, also childless, provided an education for the girl, gave her away in marriage to George Bibescu, and adopted their first-born, Grigore, who thus inherited the largest fortune and the most famous name in Wallachia.

Both brothers, Ştirbei as well as Bibescu, later became hospodars of Wallachia, one succeeding the other. Such was the reward of the father who gave his sons an education!

Among the remarkable people whom I met that evening in the home of Dashkov, I must mention also the Greek Nicolachi Mavros. He came to Wallachia in his youth, no doubt in the retinue of some Phanariot hospodar. Endowed with a remarkably fine mind and discernment, he soon made a place for himself in the political field. He served as a Turkish interpreter for our commanders in chief, Prozorovskii, Kamenskii, and Kutuzov, and received Russian ranks although he was never in actual service. He was employed especially for the purpose of bribing the Turkish military commanders and succeeded with remarkable deftness. We thus took possession of the fortress of Vidin without shedding any blood. In 1828 he was attached to General Geismar in Little Wallachia and did much to contribute to the success of our armies there. On the conclusion of the Peace of Adrianople, he was with General Kiselev and was one of the principal leaders in his administration. Kiselev put him in charge of organizing quarantines along the Danube, which was important both from a sanitary as well as a political point of view. This quarantine line alienated the Turks even more from the vassal principalities. The ambitious Mavros kept a watchful eye on his functionaries and agents and knew everything that took place on the other side of the Danube. He thus ingratiated himself with our government to such an extent that, owing to our influence, he preserved the post of inspector of the Danube Quarantines, an office which, they say, yields him a substantial income. Moreover, having put himself in a position independent of the Wallachian and Moldavian governments, he is free from control from anyone. He always played an active role in local political intrigues, but he did this so cleverly and shrewdly that he never compromised himself. He conducted himself so deftly that all parties sought his assistance. He managed his own shady affairs so well that he became one of the richest estate owners in Wallachia. His first marriage was to the sister of the hospodar, Alexander Ghica,* but a few years later, according to the prevailing custom of the region, they divorced by mutual consent so that each one could marry again. Nicolachi Mavros married Sevastiţa Sutzo and his wife married Colonel Blaramberg, a handsome and intelligent young man who served in our Engineering Corps and who transferred into Wallachian service after his marriage.

She lost nothing in this exchange because Mavros was unusually homely. However, in spite of his unfortunate appearance, he was agreeable and attractive because of his fine mind, amiability, and humor. What was most amazing about him was his phenomenal memory—not only for events, but even for names and figures. He read a great deal and he forgot nothing of what he read.

---

* He had two children from this marriage—a daughter, married to the Moldavian boyar Alecu Cantacuzino, and a son, a Russian cavalry officer. He is married to Countess Simonich, the daughter of our former ambassador in Persia.

That same evening I met also the hospodar's brothers, Mihalachi and Costachi Ghica. The former was minister of interior and the latter was *spătar*, or minister of war. Mihalachi Ghica had a bad reputation and was hated for the abuses which brought general complaints in the region. However, they had no results because of the great influence which he had over his brother, the hospodar. It can rightfully be said that he was chiefly responsible for the downfall of Prince Ghica which followed in 1842. His wife was at one time a beauty and a particular favorite of General Kiselev. Mihilachi Ghica had a son, George, who resembled his father greatly both physically and morally, and two daughters. The elder one was married to a Russian officer, Prince Koltsov-Masal'skii,* and the younger, Olga, who amazed everyone by her resemblance to General Kiselev, was married to *Beizadea* Grigore Sturdza, son of the Moldavian hospodar. Prince Ghica's second brother, *Spătar* Costachi Ghica, was a kind, honest, but worthless man. He never meddled in politics and tried hard to appear a true military figure. His wife, a famous beauty, Marițica, left him to marry Bibescu when the latter became hospodar. And Bibescu, on his part, divorced his wife, the heiress of Brâncoveanu mentioned above, making sure that her enormous fortune should not pass into someone else's hands. These examples may give an idea of the mores of the land! However, to do justice to the Wallachians, one must admit that, in spite of their easy morals, they have many admirable qualities. Their good-naturedness, hospitality, and kindness force one to forget sometimes their vices.

The following day I was introduced to the hospodar. He was a striking man. His somewhat tired face expressed passion and cunning, a combination frequently encountered in eastern types. One could not help but detect his Arnăut origin. In intellect and education he could not compare with the Moldavian hospodar, Sturdza, but he was more honest than Sturdza and thought less

---

* She soon left her handsome, kind, but soft husband, with whom she lived for a while in Petersburg. After she had left Russia permanently she took up residence in France, became involved with all kinds of men of letters there, and became a writer herself under the pen name of Dora d'Istria. She attained some recognition in this field. However, her adventures do not recommend her morally.

of his own personal advantage. In spite of this, however, his administration was poor. The abuses which he permitted aroused strong opposition in the general assembly. They were directed more against his brother Mihalachi Ghica, but since he stubbornly supported his brother, he became himself the target of general discontent. As we shall see further on, he was suspended from the post of hospodar several months after my trip to Bucharest. Before his elevation to the office of hospodar, Prince Alexander Ghica was a *spătaret*, that is, chief of the militia formed by Kiselev. He, therefore, always wore a military uniform.

We were in Bucharest only for four days, which we spent most pleasantly. I received from all sides invitations to dinners and parties. In this short time, I had the opportunity to visit, in addition to the Dashkovs, also Mavros, Bălăceanu, and the hospodar himself. I spent the most enjoyable time in the home of Bălăceanu, where there was much dancing. His daughters were justly considered to be the first beauties in Wallachia. The older one divorced her first husband and married Baron Rikman who was our consul general in Bucharest before Titov. At the time she was with him in Petersburg. Of the others who were present, one was married to *Ban* Băleanu and the other to Costachi Lenş. The rest were still unmarried. Prince Ghica arranged a musical evening, and among those who took part in the singing, Dr. Meyer attracted my attention—an exceptionally handsome and striking man with a fine voice. He wore a military dress uniform because he was the chief doctor attached to the militia. The doctor had a tremendous practice and was much respected in society. He is a brother of the well-known pianist and composer, Leopold Meyer. At this musical I also met the sisters of Colonel Blaramberg who became my close friends, just as did Dr. Meyer, during our last stay in Bucharest. As far as I can remember, the daughters of Mavros, where we had dinner, were all unmarried. The youngest, Paulina (now married to Ernest Kotsebu), was then about ten years old.

In Wallachia as well as in Moldavia practically all the old boyars still wore their national costume. The one among them who impressed me more than anyone else because of his handsome appearance and rich clothing was the *baş-boier* (the chief of the boyars—the highest distinction in the region), George Filipescu. The noble features of his face and his long, completely white beard distinguished him from the others.* Moreover, he adhered to the former times in his mode of life too. He was always surrounded by his household, and all his friends were welcome at his table. Another table was set daily for the poor on the lower floor of his home. Several dozen people gathered there sometimes. This bountiful and truly manorial hospitality so shattered his affairs that he was forced to sell his large estates one by one. However, in spite of this, he would never consent to change his mode of life. With the death of George Filipescu this honorable example from the past disappeared, and it could not be otherwise under the present conditions of life. He had two sons and two daughters. His elder son, Colonel Costache Filipescu, a well-educated man, died before his father,† and the second, George (nicknamed Dado) was educated in our Corps of Pages and, after serving several years in our Uhlan Regiment, continued his military service in Wallachia.‡ One of his daughters was married to the Moldavian Ion Ghica (called Brigadier) and the other to the Wallachian Aristid Ghica.

General Kiselev also left several Russian officers in Wallachia who had permission to enter service there. Among them I have mentioned Colonel Blaramberg. In addition, there were Colonels Gorbatskii, Poznanskii, and Viscount de Grammont, all of whom were married to local women. The latter performed with comic earnestness and purely French refinement the duties of marshal and master of ceremonies in the hospodar's court, and the first two were engaged in the organization of the militia.

The memory of General Kiselev, who lived for the greater part

---

* In the papers of the late consul general Khomchinskii, I accidentally came upon a portrait of George Filipescu. It was, I think, painted by one of the young ladies in the Blaramberg family, and the likeness was remarkable.

† He was married to a young Moldavian lady named Balş, from whom he separated on the pretext of her insanity although sufficient proof of this did not exist. To be sure she had idiosyncracies, but I could see no signs of insanity in her. Filipescu is now agent of the principalities in St. Petersburg.

‡ Dado was a great mischief maker and an odd fellow. During this brief stay in Odessa he married the daughter of a French teacher, Tricot. However, he soon left her. Certainly, she too was not distinguished by her good morals.

of the time in Bucharest, endured there even longer than in Jassy. The Wallachians spoke of him with genuine enthusiasm.

Among the members of our consulate general in Bucharest with whom I became acquainted, I recall the kind and respected Evstafii Semenovich Kotov * who managed the office for almost half a century. He used to tell in a most interesting manner of the events he had witnessed. In addition, there was also the secretary, Dendrino, a gifted but bilious and restless Greek, and Kola who held the post of second dragoman. At the consulate general also was young Durygin who came from Petersburg together with Shchulepnikov and Sokolov. They were university friends.

I also met several foreign diplomats in Bucharest. All of them were more or less envious of the exceptional position occupied by the Russian representatives. They tried to compromise our influence in the principalities, which was entirely legal at that time. The French consul general, Billecocque, and the English consul, Colquhoun, were particularly active in this respect. However, their intrigues at that time were still unsuccessful. Billecocque, a pompous Frenchman, was chargé d'affaires in Stockholm before he came to Bucharest, and he tried to give his post a political character. He was sometimes amusing in company and he outdid himself in puns, which apparently were much in vogue at that time. His English colleague was also pleasant company. He knew better than Billecocque how to conceal his political intrigues, although in reality he engaged in them with unusual zeal and even some perfidy. There was also the Austrian consul general, Timoni. He played the principal role next to the Russian representative as the spokesman of an adjacent power, that had large commercial interests and many subjects in the principalities. Timoni was a likeable and gay companion. There is no need to speak of the other consuls, because their position in the principalities was insignificant. Even Prussia limited itself to an honorary agent only (*consul honoraire*).

On the day we left Bucharest a slight frost set in which made the road better. We started out with the usual honors, accompanied by *dorobanți* with a courier ahead of us, and we drove off from the station of Urziceni in the direction of Brăila. The road

---

* He had the title of senior dragoman and had charge of current affairs.

led through the steppe and we hardly encountered a single tree. The villages on the way are few and present a lamentable picture. After three or four stops we saw the Danube on the shore of which Brăila is situated. Before reaching the city we stopped to examine a small monument in the shape of a pyramid erected in memory of the Russian soldiers who fell when taking the fortress in 1829. The siege of the fortress of Brăila, which was leveled after the Treaty of Adrianople, was led by the Grand Duke Mikhail Pavlovich, who commanded the Guard's Corps. According to rumor, this siege was not skillfully conducted. A considerable detachment was trapped in a mine dug by the Turks because of the carelessness of the august leader. The men were blown up in the air! The monument was erected on this tragic spot. Nevertheless, the fortress was taken by storm and the grand duke received the order of St. George, second grade, for this feat.

The Danube at Brăila exceeds several versts. The opposite shore, the Bulgarian, is more mountainous than the Wallachian. In general, along the length of the Danube, the right shore dominates the left, which gave a great advantage to the Turks during our wars with them. A great number of windmills testify to the flourishing grain trade in the port of Brăila, where we found several ships. The town itself seemed poor and had nothing remarkable to show. We spent but a few hours there and departed for Galatz, about two or three stations from Brăila. On the frontier, where we were met by the Moldavian authorities, we bid farewell to our Wallachian guides.

We reached Galatz at night. I recall that we heard cannon shots when still a long distance from the city. We were puzzled as to what could have been the occasion for this. We thought that revolution had broken out in the city or that the Turks had crossed the Danube. However, the explanation proved to be simple. It was the 1st of January according to the new style, and the foreign vessels anchored in the port celebrated New Year exactly at midnight.

We stopped with our consular agent, Karneev, a garrulous old man who did not give us a moment's peace. The company of his family also had nothing attractive to offer. The old man told us of his exploits in the service, but, according to our information, which Kotsebu as his immediate superior wanted to check on the spot, it appeared that things were not going well at all. This required attention, the more so because with the population increase at Galatz, where all sorts of riffraff gathered, the importance of the post occupied by Karneev, almost on the border of our frontier, was also growing.

The port of Galatz is considerably more important than that of Brăila. We found there several hundred ships which had remained to winter. The greatest part of the foreign commerce of Moldavia is carried on through Galatz, and the trade attracts here at times a large number of foreigners of all nationalities, but predominantly Greeks. The city is irregularly built,* somewhat in the Asiatic style, and the population, which already then numbered forty-five thousand, presented a great diversity of colors.

The governor, or *pîrcalab*, of Galatz was Vasile Ghica. He had common sense and business acumen. Unfortunately, however, he was not a man on whom you could depend. His wife Cleopatra, a gay and kind woman, was a Wallachian, one of the large Filipescu family.

After a two-day stay in Galatz we started out again, still through the steppes, to the city of Tecuci, which we already knew, and thence to Jassy. On the way from Galatz to Tecuci we passed an enormous estate, Pechea, which at that time belonged to Prince Moruzi. Until the war of 1829 it had not yielded a kopek of income, but with the banishment of the Turks from the left shore of the Danube, Pechea began to yield more than a hundred thousand rubles in silver in good harvest years. This is one of those substantial results of Russian patronage which the Rumanians should not forget!

We reached Jassy exactly on Christmas eve, in time for the Christmas tree which awaited us at Kotsebu's home.

I was more than pleased with this journey, which I found so interesting, and I recall with gratitude the attentions of my fellow traveler who in his relations with me was more of a good friend than a superior.

---

* Here, even more so than in Brăila, one was amazed by the countless number of windmills which present a strange spectacle.

VII
LIFE IN THE PRINCIPALITY
(1842–1845)

I CELEBRATED THE NEW YEAR of 1842 at a ball given by the elderly lady Bogdan in the company of the people whom I already knew and where everyone made a particular effort to welcome me on the occasion of my return to Jassy. The cordiality of the Moldavians as well as the Wallachians always predisposed me in their favor in spite of their many faults of which I was naturally conscious.

Two circumstances at the beginning of that year impressed themselves particularly in my memory. They were the birth of a daughter to the Kotsebus, to whom the parents gave an altogether non-German name, Zoia,* and my acquaintance with Prince Lev (Leon) Egorovich Cantacuzino, the same young man to whom I was so attracted when he appeared with Prince Alexander Mikhailovich Gorchakov in the Asiatic Department.

Prince Lev Egorovich, to whom I will refer from now on as Leon, because we became friends from the first days of our acquaintance and I never called him anything else, was the son of Prince Egor Matveevich Cantacuzino, about whom I spoke in the previous chapter. He arrived at that time from the village Bălțătești, where he lived with his family, and stopped at the home of the Kotsebus. They had become his relatives two years previously when he had married a young girl, Emilia Koskul', a sister of Kotsebu's wife, who came with her to Moldavia in 1838 when Kotsebu was appointed consul there.

Leon Cantacuzino was then no more than twenty-four years old. He was the same age as his wife, if not a little younger. Never have I met a more handsome, charming, and likable man. Everything about him was attractive—sparkling dark eyes with an in-

* Zoia in Greek means "life." They made a German diminutive, "Süsschen," from this name, which no Greek would recognize as the original name.

telligent and kind expression, fine features, a delicate complexion, and a finely outlined, small black mustache further enhanced his face. We grew so fond of each other that we became practically inseparable. Through him I came to know the aristocratic youths with whom he was popular: the princes Moruzi, his cousin, Prince Alexander Cantacuzino, and the various Sutzos, Ghicas, and others, most of whom were of Greek extraction.

The great-grandfather of the Prince Moruzi mentioned here was one of the most remarkable hospodars of Moldavia. Although his appointment came from the Porte, he devoted himself wholeheartedly to the welfare of the land which he ruled wisely and honestly. He always sympathized with his compatriots and coreligionists under the Turkish yoke. His son acted in the same spirit. He was first dragoman to the Porte, a post which carried enormous prestige in former days, but which was considered to be dangerous, because a dragoman was subject to the death penalty at the slightest displeasure or suspicion of treason. However, in those days the sultans frequently resorted to this cursory measure with all the dignitaries of his empire, including the grand vizier who sometimes paid with his life for the slightest failure. The sultans dealt in the same fashion with the hospodars of Moldavia and Wallachia, for whom the Tarpeian Cliff was not too far from the Capitol. Sometimes, quite unexpectedly, an ambassador of the sultan (*capuchehaie*) would come to Jassy or Bucharest, summon all boyars to the palace, and, after reading the terrifying firman, would put a silk noose around the neck of the condemned hospodar and strangle him on the spot. Mavrogheni, Caragea, and others lost their lives in this manner.

The first dragoman of the Porte, Prince Moruzi, aroused the sultan's displeasure because of the Peace of Bucharest in which Bessarabia was surrendered to us. After the victories near Giurgiu and Rushchuk, Kutuzov was ordered to conclude a quick peace with Turkey because Napoleon was already marching on Russia and we needed all our forces against this menacing enemy. In spite of the intrigues by the French agents, who incited the Turks against us, Kutuzov succeeded, with Moruzi's help, in concluding a profitable peace during Russia's most trying moments. This unexpected event disturbed Napoleon who depended greatly on support from this side [Turkey] in his plans against Russia. His ambassador in Constantinople used every means to provoke the sultan to a new break with us and, failing this, gave vent to his anger by turning upon the first dragoman whom he accused of treason. As a result of this, the unfortunate Moruzi was executed. Needless to say, Russia was called upon to show particular sympathy for the members of his family that he left. They were taken under Russian patronage and, therefore, able to preserve their property in Moldavia. One of the sons of this Prince Moruzi owns the estate Pechea near Galatz, mentioned in the preceding chapter. He was already in his declining years when I met him in Jassy. I often saw his two sons, Alexander and Constantin, but I must confess that they were not particularly attractive to me. Their characters and actions did not inspire me with any respect for them. Alexander married the first time the daughter of one of the elder Moldavian men of letters, Asachi, a young lady, Hermione, an intelligent and charming girl of high morals, but he left her shortly thereafter * with a child. He kidnapped from a convent a nun, by the name of Manu, with whom he entered into a second marriage. He then deserted her and married for the third time, I do not recall whom. The willfulness and cynicism of this young man, generously endowed by nature, was distasteful to me. His brother Constantin was not much better. He was married to the daughter of the boyar Nicholas Canta, and after her death married Maria Sturdza, the daughter of the lady whose adventures with Nicholas Rosnovanu I related at such length. Having squandered his fortune in card games, Constantin Moruzi launched upon all sorts of financial deals and lawsuits, the latter being his chief occupation. He lives at present in Bessarabia.

A striking contrast to these Moruzi-Pecheas were their cousins, named Zvoraştea, after the name of the lands they owned in Moldavia. They were Alexander, Panaioti, and Constantin Moruzi. The latter always lived in Greece and I, therefore, did not know him well, but I was on friendly terms with the other two, particularly

---

* She later made her permanent home in Paris where she married the famous scientist and publicist Eduard Quinet, who died recently.

with the older one, in spite of the disparity of our ages. Prince Alexander Moruzi * was very intelligent and reliable. He was particularly interested in political questions and made a thorough study of the subject. He used to come from Zvoraştea frequently, and during his stays in Jassy he would call on me daily to have a chat on political questions. These talks with so wise and enlightened a man, who was familiar with the European east, were very profitable for me. Alexander Moruzi was married to the daughter of the Moldavian boyar Răducan Rosetti. He lost her several months before our acquaintance.† He gave away all three of his daughters in marriage in Greece. His older daughter married the famous Zaimes who headed the Greek ministry several times.

Panaioti Moruzi, the friend from my early years, was also a highly educated and worthy man, but more flighty than his brother.‡

Among my new acquaintances was also Alexander Cantacuzino, a cousin of Leon Cantacuzino. He received a fairly good education in Dresden, together with his older brothers, about whom I shall have occasion to speak later, and then took up residence in Moldavia, where, when his father died, he inherited as his share an estate. He married one of the daughters of the boyar Nicholas Canta and was thus a brother-in-law of Constantin Moruzi. Alexander Canta-

---

* Although the Moruzi had no legal right to the princely title, the Russian government, having taken them under its protection in view of the tragic circumstances described by me above, never refused them the title.

† About seven or eight years later Alexander Moruzi married the widow Smărăgda Balş, a woman advanced in years and not handsome, but intelligent and rich.

‡ Panaioti Moruzi married the daughter of the boyar Michalache Paşcanu (Cantacuzino), the beautiful Aglae. They had several sons, one of whom is at present in Petersburg preparing to enter our Ministry of Foreign Affairs.

In addition to the Moruzi whom I have mentioned, there were two more members of this family who, like the Bessarabian landowners, were Russian subjects. One of them served in His Majesty's Own Hussar Regiment and was killed in 1830 during the Polish campaign. I met him several times in Tsarskoe Selo when I was in the Lyceum Boarding School. His brother, Prince Alexander Moruzi, to this day lives in Petersburg where he took up residence in his early years. As far as I know he never entered state service. Like most of the Moruzi, he was remarkably handsome in his youth.

cuzino was kind but weak-willed and completely dependent on his charming wife, Marie, who ordered him around and finally deserted him and went to live in Paris. This misfortune and his adversities he took lightly and never let them cloud his cheerful disposition. Seldom have I met a man of so happy a nature.

My first winter in Jassy was very enjoyable. I recall that toward the end of the season there was a masked ball in Kotsebu's home which proved to be a great success. I arrived in a Spanish costume. The lady I escorted was Hermione Asachi, the first wife of Prince Alexander Moruzi-Pechea.

I believe that the older brother of Leon Cantacuzino, a retired cuirassier lieutenant, Prince Grigorii Egorovich, must have come from Bessarabia not long before this ball, because I recall among the masked ladies—his wife, Maria Matveevna, and her sister, Miss Elisaveta Matveevna Krupenskii, who married a Greek officer, Dimitrachi Sutzo, about two or three years after her arrival in Moldavia. With Grigore, or Grisha, as everyone called him, Cantacuzino, I also struck up a friendship immediately, although not to the same extent as with his brother Leon to whom I was particularly drawn. He was also a handsome man, but of a different type from his brother. Tall and blond, he had the carriage and manner of a military man. Like his father he always talked with enthusiasm and lost his temper easily. Although he was then hardly thirty years old, he had experienced much in life, serving first in Bavaria, where he was trained in a military school, later in Greece, and finally in Russia. The Cuirassier Regiment of Her Imperial Highness, the Grand Duchess Elena Pavlovna, in which he served, was part of a military settlement and was located near Voznesensk. From there he used to visit his parents in Bessarabia where he married the daughter of Matei Krupenskii. The latter served for many years as vice-governor of Bessarabia. The Krupenskiis belonged to the first rank of Moldavian boyars. Matei, about whom I am speaking, had large estates in Bessarabia, became a Russian subject, and remained in this province after its annexation to Russia.* His brother, Grigore

---

* He had the rank of state counselor. His wife, the kindly Ekaterina Khristoforovna, belonged to one of the oldest Greek families, the Comnens, whose ancestors ruled in Byzantium just as did those of the Cantacuzinos.

Krupenskii, who inherited the family estates on the other side of the Prut as his share, remained a Moldavian boyar. Many other families followed the course of the Krupenskiis, as for example, the Balş, Moruzi, Donici, Sturdza, and Prunaru. That is why we have many noble estate owners with close family ties in Moldavia.

Prince Grigore Cantacuzino's wife was the older daughter of Matei Krupenskii. She was educated in Odessa and had adopted the manners of a Russian young lady. Of an extraordinarily lively temperament, even ardent, cordial and gay, Maria Matveevna was the life of our circle and often reminded me of my native country, particularly when she sang Russian folk songs and the then popular melodies. After her husband retired, the couple used to stop in Jassy on their travels. On those occasions we always met in the home of Leon Cantacuzino in Bălţăteşti.

Hardly a day ever passed here without a party and dances. We would get together in the evening and immediately send for the gypsy musicians who with their leader, the old Barbu, famous throughout Moldavia, played by ear all sorts of numbers without the slightest knowledge of how to read music. To be sure their execution was not always true to the original, and the self-taught musicians, giving full vent to their unbridled fantasies, used to wander off God knows where. However, one could never hear a false note, and their sense of timing and rhythm was so perfect that it was easier to dance to their music than to any European orchestra. They used several violins, a particular and, certainly, very ancient, instrument built of ten or more whistles, put closely next to each other—a perfect substitute for a flute—and a balalaika played by Barbu himself. It is difficult to imagine with what harmony these musicians played and how they improvised the most difficult pieces in their own way. This famous Barbu was at one time a serf in the household of the princes Cantacuzino.

Gypsy music was used for small parties, but for a large ball a

One of her sisters was married to Senator Peshchurov, the other to Katakazi, our former ambassador in Athens. The Krupenskiis had three daughters, Maria Cantacuzino, Elisaveta Suţu, and Sofia Arghiropulo. They also had two sons, Nicholas Matveevich, married to Nadezhda Ivanova Gints, and George Matveevich, married to Sofia Krasnokutskii, who died around 1860. I was on the friendliest terms with this whole family.

military band was usually engaged, excellently organized by the Austrian conductor Hefner. The musicians were for the most part gypsies and Jews, both endowed with an extraordinary gift for music.

I mingled with the young people whose leaders were the princes Cantacuzino. To this group belonged Panaioti Moruzi; Iancu (or as he was called by his diminutive name, Iancuşor) Canta with his pretty wife, Catinca; Dimitrachi Canta and his wife Pulcheria (Pulcherie); Petko Mavrogheni; Iancu Ghica; and several others. We formed a small circle and let each other know daily where we would meet in the evening. Frequently we went to the theater which then was still in the former home of Petko Mavrogheni. This modest theater was, however, comfortably built. The boxes, arranged in three tiers, were very roomy. The main floor was also well arranged. Jassy had a permanent French company, and some of the actors had real talent, as, for example, the comedian Pellier who could easily have taken his place on any Parisian stage. He was the favorite of the audience and grew so accustomed to Moldavian life that he did not want to leave the country where he enjoyed such popularity. He died in Jassy of cholera during the epidemic of 1848. How many pleasant moments did this excellent comedian give me, who at times amused me to tears!

In addition to the French company, we also had German opera, but only from time to time when some traveling company would come to Jassy. During the early part of my stay in Moldavia a fairly good company used to come to Jassy directed by Frisch, who in his devotion to art did everything on a large scale. As a result, he was soon bankrupt and went to seek his fortune in Egypt where I saw him in 1856. He traded in various European articles.

During carnival time masquerade balls were given in the same theater. They were as a rule extremely gay—men and women of the upper class, disguised as dominoes or masked, engaged in intrigues among themselves, and the people of the middle class, dressed in a variety of costumes, abandoned themselves to dancing.

It was a source of amazement to me how a population, barely emerged from the static oriental way of life, to which it was held by the Turkish domination, could adopt European manners so quickly. The cavaliers and ladies danced with dexterity. The evening always concluded with a gay and boisterous dinner. In our crowd Prince Egor Matveevich was the leader in these masquerades. In spite of his years, he participated in the most youthful manner.

Incidentally, at all these dancing parties we appeared wearing high shoes. They went out of style only in 1844 or 1845. A change in the style of clothing also occurred about the same time. The fashion was for full clothes. The style of hairdress changed also. The ridiculous vogue of tufts of hair went out of style, and men began to comb their hair back.*

After a severe winter (1842) spring arrived as it never does in our country in the north. In February we were already gathering violets during our country walks, and in March all the trees were covered with buds. I had occasion to become acquainted with the surrounding countryside of Jassy which was covered with orchards and vineyards. A favorite place for walks for the inhabitants of Jassy was Copou, a plain leading in the direction of the road to Skuliany. There during certain hours the citizenry would gather to enjoy the fresh air. Everyone who owned a carriage (and who in Jassy did not own one?) felt it his duty to take a ride to Copou. A military band, surrounded by many carriages, played there almost daily. The boyars smoked their pipes with an air of importance and the young men crowded around the ladies, who were elaborately dressed and for the most part rouged. I must say, however, that the young married and unmarried women of the upper classes did not resort to this form of embellishment. The suburb, Socola, on the other side of the city, is a much more beautiful district than Copou. However, the road there leads through a dirty street crowded with Jews and goes uphill from the gate itself. Moreover, the area is not large enough for all the carriages. Therefore, one went to Socola, not to see and be seen, but in order to walk in the magnificent and spacious garden which belonged to Prince Sturdza, in the center of which stood his country mansion. Nearby is the famous Socola seminary, the only one, I believe, in Moldavia, and somewhat further on, in the most picturesque spot is the summer house of the

---

* A daguerreotype which I happened to preserve was made in 1843, that is, before the change in our clothes. At that time a daguerreotype was still a novelty since photography had not yet been invented.

metropolitan with a garden and a vineyard. Kotsebu usually spent the summer there.

In the spring of 1842 I was forced to part with my Maksim. He had learned to speak Moldavian quickly, and had made many friends with whom he spent gay times in the coffee houses and taverns. He was never given to drinking, but he liked amusements and went to the circus often, where he succeeded in charming an equestrian performer. My Maksim was very good looking, dressed decently, and was no fool. He was popular in his own circle. I wondered often where he got the money to indulge in all the pleasures that seemed to me to be beyond his means. I had no good way of keeping track of my own money. Therefore, I never knew exactly how much I had in my purse or in the drawer. However, I began to notice that my money was disappearing a little too rapidly. I suspected Maksim and began to watch him. I was more careful and once I caught him stealing several ten-chervonets pieces from my drawer. The boy confessed and explained that the cause of it all was his infatuation with the beautiful equestrian. Sorry as I was for him, nevertheless, I could not keep him any longer. I therefore sent him for correction to Nicholas Ivanovich Kolonov in Kamenets-Podol'sk. I replaced him with a Moldavian, Todir, who spoke Russian fairly well.

Shortly after that I gave up my apartment on Main Street and moved to the consul's house. There on the lower floor (*rez de chaussée*) all the rooms had arches. In the back, in the space reserved for the archives, the cashier, deposits, etc., there were three unoccupied and fairly decent but neglected rooms. Kotsebu put them at my disposal. I remodeled them at my own expense and fixed up a comfortable place to live. I occupied this apartment during the entire period of my stay at that time in Jassy, that is, almost to the end of 1848.

At approximately this time, I met Princess Emilia Cantacuzino, the wife of Leon. She was Mme. Kotsebu's sister. Although not as beautiful as her sister, she was no less attractive in her appearance as well as in her manners, education, and, in particular, because of her extraordinarily friendly nature. She was tall and slender. Her features were delicate, and she had a very gentle expression. She worshipped her husband in whom she saw all sorts of virtues. Her trust in him was boundless. The stay of Princess Emilia Cantacuzino in Jassy this time was short. She came merely to visit her sister whom she had not seen during the entire winter because of childbirth: she had given birth to a daughter, Sofia, only a few days after the birth of Zoia Kotsebu. It was her second child. Her older one, Emil, was only a year and a half old.

With the arrival of summer the Kotsebu family moved to the dacha in Socola mentioned above. I visited them daily, and in June I took a trip of several weeks with them into the mountains. I found this excursion interesting. It afforded me the opportunity of seeing the charming sights of Upper Moldavia,* the part which borders the Austrian possessions and forms the chain of the Carpathian mountains. It may be pertinent to say here a few words about the topography of the entire region. This might be useful when following my numerous wanderings in the Danubian Principalities, and my descriptions of the various places will be more intelligible.

The entire area occupied by Moldo-Wallachia, or Rumania, as it is now called, is bordered by the chain of the Carpathian Mountains, the Danube (beginning with the rapids known by the name of the Iron Gates [portes de fer], down to its estuary), and by the Prut. These landmarks may be regarded as the natural frontiers. From the Carpathians, which reach a tremendous height and comprise such a solid mass that there are but a most insignificant number of passes along the entire length of the frontier between Wallachia and Austria, a gradual descent begins, the lower portion of which comprises an area of vineyards. Then a large plain follows which joins the Danube. The shore of this river on the Rumanian side is extremely low.

This monotonous plain, which assumes in places the aspect of a steppe, stretches over almost six hundred versts in length and a hundred to a hundred and fifty versts in width. I crossed it on my way to Bucharest, Brăila, and Galatz, as described in the previous chapter. At that time I was only familiar with but a part of the moun-

---

* The mountainous or northern part of Moldavia is called *Upper* and the rest *Lower Moldavia.*

tainous or vineyard region, namely with the districts of Jassy, Vaslui, and Bârlad. Now I had the opportunity of traveling through the mountainous zone.

Our company consisted of the Kotsebu family, Gol'tsshuer, the Lutheran pastor,* and myself. We crossed the larger boroughs, Podul-Iloaiei and Târgul-Frumos, where we changed post horses; we passed the magnificent estate Ruginoasa with its beautiful oak forest, which then belonged to the boyar Costache Sturdza. We then passed Poşcani, the estate of the vistier Nicholas Rosnovanu, and approached the river Siret where horses awaited us, sent from Bălţăteşti by Prince Cantacuzino. The unique Moldavian harness presents a handsome appearance when it is carefully made and is not like the harness of the post horses, where scraggly small horses are tied with plain thin ropes and the *surugii* are almost always in rags. Imitating the boyars in these regions, the princes Cantacuzino liked to show off their horses, and their *surugii*, dressed in elegant national costumes, made a pretty picture astride the horses, driving them with great agility. In spite of the mountainous region and the hillsides, we drove quite fast and I marveled at the care of the *surugi* who could anticipate the slightest bumps. Particular adroitness and skill were required on the part of the *surugi* nearest the shafts.

The longer we drove the more picturesque the scenery grew. Beyond the river Siret we no longer encountered vineyards. Even wheat appeared more seldom, and, finally, it was replaced throughout by rye. Only here and there could fields of maize be seen. Everything pointed to a more severe climate—the result of a higher region. A magnificent picture opened up before us as we approached the district city of Neamţ, situated in a beautiful valley, girdled by mountains. The city itself was not remarkable. I could not see a single outstanding building, almost nothing but *mazankii* [whitewashed clay cottages] on the order of our Ukranian ones, but not as neat because of the great number of Jews who do not like cleanliness.

* There is a small Lutheran church in Jassy built on the grave of one of the Russian generals, whose name I cannot recall. It was, therefore, under our patronage. Afterward, the church passed under the jurisdiction of the Prussian consul.

All the way from Jassy we met camps of gypsies who spend practically all their lives under the open sky and who roam from place to place. In winter they build dugouts, from which they crawl with the first rays of sunshine. I was stunned by their savage appearance. Their swarthy faces were not devoid of beauty, but the tousled black hair, never touched by a comb, and the absence of any kind of clothing, except for a dirty torn shirt, present a disagreeable sight. Their women grow old very early, and the most beautiful gypsy girl at thirty turns into a loathsome old woman. The stark naked children raise a terrible cry on the appearance of a passer-by or a carriage and pursue them with entreaties for alms. The seminude women do the same, running sometimes till exhaustion after a carriage with a breast-fed infant in arms. These nomadic people live in incredible poverty and filth. The only trade they engage in is that of the blacksmith. Almost all blacksmiths in Moldavia are gypsies. However, they live chiefly on thievery. In spite of all the precautions taken throughout the neighborhood where gypsies make their camp, it is almost impossible to guard against these nimble thieves. In the event of a considerable robbery near a gypsy camp, the government officials usually go to the chief of the camp. He will always be willing to bear a light punishment rather than surrender the object stolen, but if threatened with severe punishment, he can always produce it. These chiefs, like the atamans living among robbers on the stage, wear a long red cape as a mark of distinction.

I am speaking now of the nomadic gypsies, but there are gypsies who lead a settled life in villages and towns. They are domestics for the most part—lackeys, coachmen, and, in particular, cooks. They are quite agile and intelligent, but exceedingly slovenly. I have already spoken of their remarkable gift for music. In practically every village you will find a lăutar, as a gypsy musician is called. No village holiday is complete without him, and in the cities there are assemblies of lăutari. The most remarkable of them was that of the old man, Barbu, to whose music we enjoyed dancing so much in Jassy.

Until 1844 or 1845 all gypsies were in a state of serfdom. The princes Cantacuzino owned several thousand of them. A large number of gypsies were registered as chattels of the monasteries. They were freed for an insignificant redemption payment, the owners getting but a small part of it. The reform was passed peacefully and with no complaining on the part of the owners. To be sure, these serfs were practically no source of income to them.

From Neamț it took us no more than half an hour to reach Bălțătești, where we stopped for several days.

I cannot find words to express the pleasing impression made upon me by this charming estate. Bălțătești is as picturesque as it is cheerful, situated between beautiful rolling hills among which are scattered small peasant houses, *mazanki,* surrounded with gardens. Here the mountains, which are seen on the horizon, do not obstruct the range of vision and merely serve as an embellishment of the general picture. The spacious manor house with its grounds stood on an elevation. A steep path led to it. Our *surugii* dashed up it with the usual drawn-out yell and the deafening cracking of their long whips. The Cantacuzinos met us on the porch with open arms. One could not help but feel a particular pleasure in the company of this friendly, cheerful, and kind family, upon whom fortune smiled so much at that time.

Since we had to travel further into the mountains, we stopped in Bălțătești only for a day and a night this time. Our final goal was Borca, a settlement which, as far as I can remember, belonged to the monastery Slatina, situated high in the Carpathians. Our road led through Hangu, the largest of the Cantacuzino's estates, which comprised thirty-five thousand desiatins.[1] We went there accompanied by Prince Leon, who took upon himself all the details connected with our difficult voyage.

Although Bălțătești is adjacent to Hangu—it too extends over several thousand desiatins (I think five thousand)—we were faced with a journey of twelve hours to reach the manor house in Hangu where we were to spend the night. On leaving Bălțătești we began to ascend the Děalul Doamnei (the Lord's Mountain). Here nature assumes a more severe character—the deciduous trees gradually disappear and are finally almost completely replaced by pine. Only here and there does one encounter our Russian birch. The fields also become less cultivated. First maize, then rye disappear, nor can any barley be seen either. Potatoes and oats are encountered only rarely. On the other hand, the mountains, covered by dense mast timber, rise ever higher, and the road grows narrower. Before entering

Hangu a clearing opens up. The view of the settlement is charming. The pretty frame houses of Hangu are scattered along the shores of the noisy rushing waters of the river Bistriţa, which is used to float timber. Everywhere grazing herds of sheep can be seen, and rising on the horizon is the gigantic mountain Ciahlău, its peak almost always veiled in mist. It is from seven to eight thousand feet high. We were met by a large group of riders astride small but strong mountain horses. All were armed with shot guns. The Hangu men never part with their shot guns when on the road in this magnificent land which abounds in a variety of game. As we approached the settlement, the number of riders grew until it finally reached several hundred. We thus made our entrance escorted by a sizable detachment. When we ascended from the carriage, the riders, or *plăiaşi*, as they are called there, fired a salute in our honor.

The Hangu natives are distinguished by their tall stature and their remarkable physique. They wear their hair long, practically down to their shoulders. On their feet they wear artfully woven bark shoes in which they walk easily over the forest-covered mountains, where the passes are difficult because of the obstructions encountered at each step, so that no boots, no matter how strong, would last longer than a day. All are excellent hunters. It is a rare Hangu man who has had no occasion to struggle with a bear whom he faces with unusual courage. I doubt whether there is as much game anywhere else, especially wild goats. Prince Leon, who participated frequently in these hunting expeditions, acquired such skill that he delighted even the Hangu men. They were so fond of him, as well as of his entire family, and so devoted to him that they would go through fire and water for him.

Hangu consists of five or six settlements, numbering more than a thousand people. Agriculture is practically nonexistent and, naturally, could not be carried on there. The entire wealth of the area is the forests which abound in magnificent mast timber. Each such pine tree (*catarg*) of the better quality at that time used to cost in Galatz ten or more chervonets, that is, more than thirty rubles in silver. The prices today, of course, have doubled. From Hangu the timber was floated down the Bistriţa and the Siret to the Danube, and from there it was distributed all over Europe. This timber is used also for large rafts which float rapidly down the current. The raft is ordinarily guided by two peasants—one stands at the rudder and the other manipulates a large hook to prevent collisions with other rafts, and also to push off from the shore.

We settled down for the night in the home of the tenant or lease-holder, a Hungarian who entertained us excellently. Toward morning we were cold, despite the clear weather, and did not know how to get warm. And this was in the month of July! The difference in climate between our Upper Moldavia and this region is quite notable. Only a few days ago in Jassy we endured terrible heat and now we had to bundle ourselves up because we were so cold!

On the following day we set out further. We were faced with about ten hours more of travel before reaching Borca. We passed through the several settlements comprising Hangu, accompanied by many riders. Not for a moment did we lose sight of majestic Ciahlău and the river Bistriţa, along whose shores several sawmills were busily working. We crossed the river several times. Thereafter, we entered a canyon which we did not leave until we had almost reached Borca. By evening it had grown so dark that we could barely move ahead, and even then thanks only to the riders who rode ahead of us with torches. I can imagine how our appearance must have horrified the fur-bearing inhabitants of these vast forests! We heard all the time the roaring of waterfalls, and finally the Bistriţa appeared again. We entered a large clearing and here lights flashed in the few small houses which comprised the entire settlement of Borca.

After so tiring a trip we hoped to enjoy a rest on soft divans and in armchairs, but alas! Wooden benches and tables comprised the only furniture of the modest and spotlessly clean quarters prepared for us. Fortunately, the van which followed us from Bălţăteşti was loaded, thanks to the kindness and foresight of the Cantacuzinos, with enough mattresses and pillows, which rendered us excellent service during our three-day stay in Borca. As for our food in this remote spot where it was impossible to obtain anything, the monasteries Slatina and Neamţ took care of it for us. Although far from us, they sent their *plăiaşi* with provisions. As a result we had more wild goat meat than we knew what to do with, in spite of the ferocious appetites which the fresh mountain air and walking stimulated in us.

Borca is situated in a majestic, but wild area. I do not know how we ever got there in a carriage over a road suitable only for

pedestrians and people on horseback. The total peasant population
of Borca consisted of ten peasants engaged in felling trees. In the
surrounding area at a distance of two or three hours of travel, there
is not a single dwelling. However, it was just this remote corner that
we liked, among the gigantic mountains clothed in primeval forests, with this roar of countless cascades and this magnificent
Bistriţa which in its bends and turns suddenly appears where you
do not expect it and amazes you with its rapid current and the
clarity of its waters.

However, we were not alone in this seclusion. Some of our
friends from Jassy visited us. Among them I recall Iancuşor Canta,
who came with his sister, Maria Cantacuzino (Prince Alexander
Cantacuzino's wife), and her friend, Hermione Asachi, mentioned
earlier. Both these charming and highly educated ladies enlivened
our company. Their dreamy and poetic mood found abundant
nourishment in the natural scenery surrounding us. Their visit was
followed by one from Costache Sutzo, the son of the *beizadea*
Nicolachi Sutzo, the most intelligent and capable of Moldavian
statesmen, and Dr. Frenkel. One day we took a wonderful trip
into the mountains with these men. We heard from the peasants
that there was a magnificent waterfall two or three hours from
Borca, but that the road was almost impassable. We got together
some food, took a few guides with us, and started out. We encountered obstacles at practically every step—either a fallen tree,
or a brook, or tremendous boulders, or a stump, or a thicket so
tightly interwoven in the dense forest that we could barely pass
through it, and then only after the guides had chopped a way past
for us with their axes. We were climbing, which tired us considerably. Finally, we reached the cascade. Here we forgot our weariness
and were carried away in admiration of the heavenly panorama.
The water fell in a wide sheet from a great height, foaming down
with a roar which we heard at least a verst before reaching the
cascade. The whole setting is beautiful and wonderful in its wild-
ness. Rarely does a traveler reach this God-forsaken place. The
guides could not recall any boyar, that is, "gentleman" in Moldavian, who had visited this region. I cannot recall the name of this
waterfall.

We walked about twelve hours and returned exhausted and completely in rags. Soon after that we left Borca and rode to Hangu. From there the Kotsebu family proceeded on their journey to Bălțătești, and Dr. Frenkel, Sutzo, and I remained for another twenty-four hours in Hangu to ascend the mountain Ciahlău. Leon Cantacuzino, who came to meet us, remained with us, as did Shchulepnikov, who brought the mail from Jassy.

The weather favored us. There was not a cloud on the peak of Ciahlău, which we reached after a three-hour difficult ascent. The view from there over the surrounding area is magnificent. The clouds appeared to wait till we had enough of enjoying the beautiful view and had begun our descent down the mountain. The sky above us was clear and the sun was shining brightly. However, the clouds below us were so thick that we could see nothing below. As we came down further we found ourselves surrounded by mist, and at the foot of the mountain we were caught in a heavy rain. This phenomenon seemed quite amazing to us.

Fortunately, it soon cleared up, and our gay party began to make preparations to continue on its way. From the time we parted company with the Kotsebu family we had neither cook nor provisions. We had to be content with what food we could get from the peasants: eggs, *brânză,* a sort of cheese made of goat's milk, and *mămăligă* which replaced bread among the common people of Moldavia. This is merely a hard-cooked gruel of maize, used also in Italy and known there as *polenta.* They usually eat it with fresh butter, and the Moldavians like it so much that the dish is often served at the table of the boyars. We found wine everywhere. In Moldavia it is as common as *kvas* is in Russia. However, naturally the wine used by the common people is of poor quality and is perhaps as sour as our *kvas.*

We decided to go down the river Bistrița to the city of Piatra and to take a carriage from there to Bălțătești, which is no more than three hours' ride from Piatra.

A raft of simple construction was made ready for us. This was nothing more than several logs of great size firmly attached together, with no sides or rails. One enormous oar was attached to the front and one in the back, and the raft was operated this way. The men who use these oars must act in unison. Therefore, the one who is in the front takes orders from the man at the rear. The only moving power is the current of the river which at times is so swift that the raft rushes faster than any steamer. Our raft was very roomy and was decorated with young green branches as is customary with us on Whitsunday. Grass scattered over the floor served us in place of a rug. Everything smelled fresh. One of the helmsmen began to play a reed pipe and another broke into song and we sailed a whole day to the accompaniment of this melancholy but pleasant music. The shores of the Bistriţa are unusually picturesque. Once we had passed the section of dense forests we began to see now on one, now on the other shore of the river beautiful villages and rich pastures where herds of sheep and, occasionally, horned cattle were grazing. Later wheatfields began to appear. All this presented a beautiful picture. However, never have I seen so lovely and gay a sight as that which the small district town, Piatra, presented, which we entered as the sun was setting. The houses were scattered—some along the shore of the river, others on the slopes of the mountain, and still others in the valley. And all of them were in greenery. The population of Piatra numbered no more than three to four thousand.

We stopped at the home of the district police officer, Costache Ghica, a hospitable man, who did not, however, enjoy a good reputation morally. He had paid a high price for his post to the authorities, among them also, no doubt, to the hospodar, with whom everything was for sale, and who therefore took bribes from his subordinates mercilessly. His wife, Catinca, was the daughter of the old boyar, Catargiu, who still wore the national costume and fulfilled at that time the function of *aga*, or chief of police of the city of Jassy, a post held in great esteem in Moldavia, no doubt because it was very profitable.*

We stayed overnight at the home of Costache Ghica and left the following morning for Bălţăteşti where we found quite a company.

* This Catargiu had, in addition to his daughter Catinca, five or six sons. The youngest, Lascar, turned out to be a remarkable statesman. From the time of the merging of the two principalities, he played a prominent role, heading the well-meaning liberals, who rightly referred to themselves as conservatives. Lascar Catargiu has been for many years now the first minister, in spite of the intrigues of the radicals who tried to eliminate him.

I spent at least ten days more there engaged in varied diversions and magnificent trips.

Sometimes we would send to Piatra for *lăutari* and hold a dance; sometimes we went hunting, but most frequently we took excursions to neighboring monasteries.

Once on a holiday the *lăutari* arrived in the morning. The Cantacuzinos thought that it would be a good idea to have a village celebration and they called together the entire village. The *țărani* and *țărance* (men and women peasants) took each other's hands and formed a large circle, with the *lăutari* in the center. The circle began to move, now one way, now the other. First they moved slowly and at a given moment all stamped one foot in time. They then moved faster, and stamped their feet harder. The musicians played faster and faster and the dance continued until exhaustion. This national dance is called the *horă*. They say that it goes back to the days of paganism, which I readily believe because it actually does have a savage character. Then the *țărani* danced in pairs some less original dance, whose name I cannot recall. It somewhat resembles the Hungarian *csárdás*. The holiday costumes of the Moldavian *țărani* resemble our Ukranian costume very much, except that they like to trim their wide-brimmed summer felt hats with ribbons of various colors. The girls are fond of beads and necklaces made of coins. They are quite handsome, but I have encountered no beauties among them. The Transylvanians are the most beautiful of the Rumanian types. The reason for this is perhaps the fact that they have mixed with the Hungarian race.

I found hunting on the outskirts of Bălțătești very amusing. I experienced at that time this pleasure for the first time in my life, and my joy knew no bounds when I was lucky enough to kill a grouse. However, I did not become a true hunter and I never kept a shot gun.

The trips to the monasteries, however, were of the greatest interest to me. A nunnery, *Văratic*, is no more than twenty minutes drive from Bălțătești. There we were welcomed with the ringing of bells and were led directly to the church, which stood in the center of a large courtyard, around which were buildings, including the convent, with individual cells for the nuns. The Mother Superior came to greet us accompanied by many nuns who followed her in pairs. She led us into the church where a short service was conducted in our honor. The architecture and the interior of the church are the same as that of all Orthodox churches in the East. They differ from the Russian in having always seats along the wall with high arms that one can lean upon, but only when standing; after that the seat springs up and clamps tightly against the wall. These seats, or stalls, elegantly carved of wood, make the standing of the worshippers easier during the long services of the Orthodox Church. I cannot understand why they have not been adopted in Russia. The churches in Moldavia and throughout the East differ also from ours in that no images of contemporary art are permitted. The faces of the saints are homely and are painted entirely in dark colors. Apparently every effort is made so that they will not bear any resemblance to real people. Our Suzdal' paintings are viewed as the height of perfection by this primitive school! [2] The church singing in the East is unattractive too. It is drawn out and nasal. A natural voice is not permitted. On leaving the church the Mother Superior escorted us to the refectory and then to the individual cells. We saw everywhere simplicity, comfort, and immaculate cleanliness. I was surprised by the noble features and fine manner of many of the nuns. Some of them spoke French fluently. It appeared that they belonged to some of the best boyar families. The Mother Superior was herself a member of the Sturdza family.

Among the first-rank Moldavian families in those days the custom prevailed of dedicating the youngest in the family to a monastic order, just as was once true in France. This was not prompted by any religious motive, but to ensure his future from an economic standpoint. Majority or primogeniture never existed in the principalities. An inheritance was divided equally among all children, irrespective of sex. The dedication of one of the children to a monastic order was done in order to augment the share of the others. I say "children" because they were sent away to a religious order at a tender age, sometimes as young as twelve or even ten. The sad consequences of such a scandalous custom are obvious. Children, who felt no calling to monastic life and who realized later the injustice done them by forcing them into it, often sought consolation not in prayers but in sensual pleasures, and finally with the years led a depraved life. Such was the fate of most of these unfortunate creatures, particularly the women who had practically no future. The fate of the men was somewhat better. With some education and other qualifications they could hope for a brilliant career in the ecclesiastical field. The heads of the fabulously rich monasteries, the bishops, archbishops, and, finally, the metropolitan, who had a tremendous income and an important position, were selected almost entirely from the monks who belonged to the best families in Moldavia. These ecclesiastical dignitaries exerted a great influence even in politics because the metropolitan was a representative, and the eparchial bishops were the first members, of the national legislative assembly.

We visited another convent, Agapia, about forty minutes' ride from Bălţăteşti. Agapia is situated in a charming valley surrounded with high mountains and covered with a dense forest. The organization of this convent is the same as that of Văratic. We met here many beauties among the young nuns. The buildings of the convent were crowded with visitors from Jassy—young men, for the most part, whose free and easy manner with the nuns gave me a most unfavorable opinion on the severity of morals in this holy retreat.

At that time the Moldavian nuns, in general, enjoyed much freedom. They were allowed to visit relatives, and they sometimes absented themselves from the convent for several days under this pretext. Many of them visited their neighbors, the Cantacuzinos. The old *mamă* (mother) Cuza from Văratic used to come practically every day to Bălţăteşti and would spend the whole day there. We had a great deal of fun at the expense of this kind but limited and superstitious woman. What tricks would we not invent to amuse ourselves at her expense! She believed everything with a truly childish simple-mindedness! She was greatly attached to the family of the Cantacuzinos.

The most remarkable monastery in Moldavia is, of course, Neamţ, situated a few versts from the city of the same name. We were received there with unusual honors. The entire fraternity came out to greet us, headed by the clergy in full vestment, followed by the monks in pairs, in their robes, holding long candles, and behind them the rest of the brotherhood. We entered the church to the accompaniment of a deafening ringing of the church bells where a service was given in Russian with the assistance of two choirs, one Russian and the other Moldavian. The monastery of Neamṭ had up to a thousand monks, of whom more than one hundred were Russian. At that time they were accepted there gladly, particularly those who had good voices. The statute of the monastery itself was completely Russian in character. It was introduced during the past century by Father Paisii, a most unusual man in mind, character, and Christian mode of life. Paisii was Russian by birth. However, in spite of his foreign origin, he acquired so much influence in the monastery that he was elected Father Superior. The period of his administration is regarded as the most brilliant in the history of the monastery of Neamṭ. His memory is honored there almost as much as that of the holy fathers of the early centuries of Christianity.

After the service we went to the Father Superior, or the Elder of the Monastery, as he is called in Moldavia, Archimandrite Neonil, who, suffering from paralysis, has not left his bed for several years. In spite of his sufferings, Father Neonil was very active and energetic in his management of the monastery, which at that time had an enormous income at its disposal. In addition to large estates in Moldavia, Neamṭ also owned several villages in Bessarabia. The immediate assistant of Father Neonil, but only in spiritual matters, was a Russian *protomonakh*, Nafanail, a kind and venerable man, whom I got to know. Working under him was a young deacon, Augustin, also of Russian stock, who had served somewhere in the civil administration in Russia before taking his monastic vows. Thanks to these compatriots, who always received me with great hospitality, I spent many pleasant moments in the beautiful monastery of Neamṭ, which I visited so frequently during my stay of many years in the Danubian Principalities.

We examined the vestry and depository of the monastery where much precious church plate of great historic value was preserved. I recall among them a New Testament and an altar cross presented by Peter the Great and Empress Catherine II. I also saw preserved there some historic acts, annals, and other manuscripts of the most ancient times. The monastery of Neamṭ always played an important role in the history of Moldavia. It was often besieged by the enemy during the wars with the neighboring people, and one of the most popular of Moldavia's rulers, Stefan the Great, gained fame by its walls. In this respect Neamţ reminds one somewhat of our own Holy Trinity–St. Sergius monastery.

The monastery is situated in a beautiful and picturesque valley. A magnificent oak forest is next to it. The entire place is charming and the climate is excellent. The monks live a free and easy life here and abandon themselves to the contemplative life. Unfortunately, many of them are not satisfied with spiritual joys. Religious feeling, which one might expect would find so much nourishment in the contemplation of this wonderful natural scenery, is not developed in them, and idleness breeds in them material and carnal inclinations. The proximity of the convents presents a great temptation. The kind and worthy metropolitan of Moldavia, Veniamin, wishing to put an end to the criminal relations of the monks with the nuns, pronounced a solemn curse upon the path between the monastery and the convents, trampled down by the steps of these adulterers. But the measure proved to be inadequate. The monks broke another path, and the honorable pastor once received the mournful news that, in cleaning a pond near one of the convents, several remains of newly born infants were found! While speaking of the immorality of the Moldavian monks, one should add, nevertheless, that there are very worthy people among them. Unfortunately, however, they are an exception to the rule.

I must also do justice to the hospitality of the Moldavian monasteries. In the spacious building of the Neamţ cloister, large rooms were reserved for worshippers of all classes as well as for all kinds of travelers. Even foreigners were made welcome there. Many lived there for weeks. They were denied nothing, and no requests for contributions were made of them. Moreover, the monastery, in view of its great means, was in no need of this.

My first trip into Upper Moldavia left the most pleasant memories with me. We returned to Jassy by the same route. On entering the zone of vineyards near Târgul-Frumos, I noticed with satisfaction that the grapes were beginning to ripen. Never yet had I had the opportunity to see grapes in such abundance. Practically all vineyards in Moldavia are open to any one passing by. You can crawl into them and eat the fruit to your heart's content—not like western Europe where you are dragged to a police station for picking a vine. I must admit that this broad freedom, unhampered by the stinginess of the owners, is much to my liking. The truth of the matter is that with the abundance of grapes the loss from taking such liberties is insignificant.

The fall is the best time of the year in Moldavia. The constantly clear weather and the pleasantest temperature continue down to the middle and sometimes to the end of November. We took walks daily along the beautiful suburbs of Jassy, which abounded in vineyards and orchards. To me, a native of the North, this manifestation of nature seemed marvelous at a time when the most inclement weather prevails in Petersburg. The cheapness of the fruit in Moldavia amazed me. You could buy the best watermelon or musk melon for a few kopeks, not to mention apples, pears, plums, and cherries. But all is not good which gives pleasure. With the appearance of the fruit, an epidemic of fever begins in Moldavia, which spreads so much that at times there is not a household where someone is not suffering from this disease. Although Moldavian intermittent fever is not dangerous, it is extremely painful. The paroxysms are generally repeated every other day. They begin with chills which turn to a temperature with terrible headaches and end with profuse perspiration. The patient also vomits, has diarrhea, and reaches a state of complete exhaustion. The unpleasant part about this illness is that it frequently recurs. These relapses last sometimes an entire year, and even longer. You think that you are rid of the disease when suddenly, at the slightest carelessness, it recurs with the former force. This happened to me in 1853. Until that time I had not had a single attack of fever although I was not at all careful. Spring fevers also occur, but they are not as stubborn. There is no doubt that fruit, and particularly unripe fruit, merely aggravates, but does not engender the epidemic. Its cause must be sought in the poor sanitary conditions of the locality.

Upon returning to Jassy I again took up my work in the consulate. When I was sufficiently familiar with the business correspondence, I began to write readily papers in Russian to our authorities as well as memoranda in French to the office of the *postelnic* and the foreign consulates in Jassy. Our heaviest correspondence was with the Austrian and Greek consulates who had the largest number of subjects in Moldavia. We had under our patronage no more than two or three hundred persons despite the proximity of the frontier because we recognized as Russian subjects only those persons who carried official passports or other identification. The other consulates, however, abusing their authority, issued passports [of the countries they represented] not only to people who had no right to them but even to natives and fugitives of all possible nationalities. This was done purely from mercenary motives. They charged a chervonet for each passport they issued, and this was a considerable source of income for them. As for our Russian fugitives and deserters, and those were numerous under the conditions of serfdom and the order then generally prevailing, on crossing the Prut they became subjects of some foreign power thanks to the mercenary consuls. I had occasion to see our bearded Great Russians with German, Greek, and even French or English passports! These scoundrels had disputes from time to time with our subjects, and we were then forced on account of these former Russians to engage in official correspondence with some consulate which treated our fugitives either as French, Greek, or natives of God knows what country. I can say emphatically that only the Russian agents conducted themselves honestly and did not allow themselves such crying abuses. It should also be added that every consul (except the Russian) appointed an agent, or an elder, as he was called there, in each district town, and these secondary agents, emulating their superiors, created for themselves as many subjects as they pleased. One can well imagine the evil consequences of this in a land where consular jurisdiction exists on the basis of the capitulations with Turkey! The local administration could not resort to the simplest police measure without encountering the consul's interference.

General Kiselev tried to put an end to this crying abuse by inserting a special article in the Organic Statute to the effect that only persons equipped with official passports of their native country are recognized as foreign subjects. According to this article also consulates were forbidden to appoint elders to represent them in district towns. However, this fair order was carried out only by the Russian consulates. All others refused to comply with it. They did not accept the laws introduced in the principalities by the Organic Statute, even when these pertained only to internal administration. They continued to regard the principalities as Turkish provinces where the ancient capitulations concluded with the Porte were in full force. This view was supported by all Western powers, no doubt from envy of the favorable position in the principalities acquired by Russia at the cost of her blood, and in a place where the majority of the population belonged to the same Greek Orthodox Church. Moreover, in his desire to root out the abuses mentioned above, General Kiselev did not eliminate consular jurisdiction based on the treaties or the capitulations. He merely tried to eliminate their evil and incorrect application.

In time, when the Western powers began to take an interest in the fortunes of the principalities and when they assumed the protectorship over the principalities, they began to watch more closely the behavior of their agents, and the abuses under discussion began to decrease. Their first step was to pay closer attention to the selection of consuls, which down to the 'forties had been extremely poor. Austria was an exception in this respect, but England and particularly Prussia, not to mention the secondary powers, appointed the most unworthy people to the posts of consuls.

In the early years of my stay in Moldavia the Austrian consul general was one Eisenbach, formerly postmaster at Constantinople. He was a kind, but quite limited man. Attached to him as secretary was first, V. von Starkenfils and later Isfording, both honest and well educated young men, particularly the latter. The French consul was one Hubert, quite an amusing old man, who, as I heard, was formerly a tutor in a Russian family, and the English consul was Samuel Gardner, who for some reason had also lived formerly for a long time in southern Russia and suffered greatly from spleen. Both of them paid little attention to their official functions. Instead they relegated them to their dragomans, men who were most dishonest. However, nowhere were there as many abuses as in the Prussian consulate which was in charge of a Jew, Kalman [Kahlmann]. Under him the number of Prussian subjects in Moldavia increased to several thousand, of whom hardly any one had ever been in Prussia. All of them were Jews or our Russian fugitives who bought passports from Kalman for themselves. Complaints

against this scoundrel finally reached Berlin, and he was replaced in 1843 by *Justiz-Geheimrath* Neigebauer, who was given
the title of consul general for both principalities. The calculating
Prussian government did not assign him a salary, but conceded to
him the right to use the consulate's income. Neigebauer, therefore, found it unprofitable to decrease the number of the subjects
under his jurisdiction. Everything moved along as usual, although,
of course, the abuses were fewer than under the Jew Kalman.

Seldom have I met such an eccentric as Neigebauer. He was
fond of company and wanted to appear a man of the world, which
was amusing considering his comic appearance, awkwardness, and
poor French pronunciation. In spite of his age, he was around sixty,
and a homely figure, which was made even uglier by a monstrous
wig, he laid claim to success with the opposite sex and chased the
Moldavian women who always made fun of him. Moreover, he
tried to appear as a learned man, and one always saw him with a
book in hand, engrossed in reading even if he was riding in a carriage or was on horseback. He rode on one of the horses of his
carriage, a real jade, and his coachman rode astride the other as a
sort of jockey. In this fashion he once rode past the theater, in front
of which hung on a rope strung across the narrow street a small
billboard announcing that there would be no performance that
evening. With his nose buried in a book, the poor rider did not
notice the rope and was caught around the neck by it; he rolled
off his horse. Neigebauer used to visit Kotsebu almost daily, and
I must admit that we could not deny ourselves the pleasure of
ridiculing him. At the time he was busy writing a book on Moldavia. When collecting data on the region, he bored us to death
with his questions. He had the habit of putting down our remarks
in his notebook. Once he approached my colleague, Shchulepnikov,
and asked him if he could give him some information on the
customs and habits of the so-called *Lipovani*, a Russian population
living in considerable number in Moldavia. To play a trick on
Neigebauer, Shchulepnikov began to tell him fantastic stories about
the mode of life of these people and without hesitation said to him
whatever nonsense came into his head. Neigebauer listened to him
carefully all evening, eagerly recording the lies.[3] I had no occasion to see subsequently Neigebauer's published book on Moldavia,

and, consequently, cannot be sure that Shchulepnikov's joke found its place there.

At the end of 1842 we had to part company with the kind Sokolov. He was appointed secretary in the newly established consulate in Galatz and remained there several years under consul Kola, from whom he endured a great deal. Kola was an honest man and an exemplary official, but not talented and hard to get along with. In Galatz he married the daughter of our honored diplomat, Privy Counselor Fanton. She was the sister-in-law of my uncle Teofil Petrovich Pankrat'ev. The old Karneev, the former agent in Galatz, about whom I spoke in the previous chapter, was relieved from office.

Approximately around this time, Major General Alexander Osipovich Diugamel' [Duhamel] of the retinue of his Imperial Majesty came to us from Petersburg. He had just returned from Persia, where he had served as ambassador for many years. He was now going to Bucharest in the capacity of commissioner, along with the Turkish commissioner (I think it was Shekib Effendi) to investigate complaints against the hospodar, Alexander Ghica, presented to both courts by the Wallachian General Assembly.

I already mentioned the dissatisfaction of the people of Wallachia with the hospodar because of his poor administration and the abuses of his brother, Mihalache Ghica, who was then minister of interior. Naturally schemes were contrived here by such persons as Ştirbei, Bibescu, and Iordache Filipescu, who aspired to the princely position. Each had his own party, and all of them joined together to overthrow Ghica. A complaint was drawn up at the General Assembly of Deputies (*Assemblée Générale*), enumerating the abuses of Prince Ghica's administration, and it was resolved by an overwhelming majority of votes that the complaint be submitted to both courts. As a consequence of these *doléances*, as they were called in diplomatic language, the Russian and Turkish governments appointed the mentioned commissioners to investigate the case. One can imagine the intrigues that brought about the appearance of these political inspectors. The Turkish commissioner, as is customary, took full advantage of the situation, and made a fortune by taking bribes from the hospodar as well as from each of the candidates for the office of prince by promising each individual the support of the Porte. While our representative, listening to anybody and everybody, tried to form a fair opinion on the state of affairs, he was deceived, being unfamiliar with both the region and the people. Be it as it may, the reports of the commissioners were not in favor of Hospodar Ghica. Both courts decided to relieve him of the office. This was done in the spring of 1843 by a firman of the sultan, which was brought to and officially read in Bucharest by a special Turkish commissioner who in his turn filled his pockets handsomely by accepting money from the provisional government (Caimacamie) and from all the boyars who sought the office of prince.

The fall of Alexander Ghica was a great victory for Ia. A. Dashkov, who had always criticized the administration of the hospodar in his reports. There could be no doubt but that this administration was bad. But could anything better be expected of his opponents, trained in the school of political intrigue, as were all the boyars of that time—intrigue of which you can receive a true conception only in the East? In this respect the kind and noble Iakov Andreevich was, it seems to me, mistaken and fell too much under the influence of the ambitious George Bibescu with whom he was on friendly terms. Moreover, Ghica's fall, having demonstrated the instability of the hospodar's throne, increased the spirit of opposition which always distinguished the boyars and gave each the hope of sitting on the throne.

The Wallachian Extraordinary General Assembly, created in accordance with the Organic Statute, elected George Bibescu as hospodar by a considerable majority of votes. His most dangerous opponent was his own brother, Barbu Ştirbei. Thanks to the influence of Ia. A. Dashkov, the partisans of the third candidate, the old George Filipescu, whom I mentioned before, realizing that they would be unable to put through their candidate, gave their support to Bibescu and thus gave him the majority over his older brother.

The dethroned Ghica retired to Naples and took up residence near his good friend, Countess Sukhtelen, with whom he had become intimate during the Turkish campaign of 1828 and the occupation of the principalities by our troops. At that time she lived in Bucharest with her husband, a Russian cavalry general.* From Naples Ghica followed through his agents everything that took place in Wallachia and, by his advice and particularly with his money, assisted greatly in the formation of the revolutionary party which in 1848 dethroned his enemy and successor, Bibescu, and at the same time was responsible for so greatly undermining our influence in the Rumanian principalities. Ia. A. Dashkov, who in all other circumstances proved to be a wise diplomat, did not, of course, foresee these consequences. In my inexperience, I was at that time an admirer of the actions of my chief whom I sincerely loved and respected.

The revolution that took place in Wallachia naturally had repercussions in Moldavia. The boyars were aroused there also. The party hostile to Prince Sturdza tried to obtain enough votes in the General Assembly to draw up an act exposing the hospodar. His sins were even greater than those of Prince Ghica. However, it was not easy to deal with the shrewd and intelligent Sturdza. He managed his affairs so cleverly that instead of complaints the chamber presented him with a citation of appreciation in which it referred to him as the "Father of the Fatherland."

Both Moldavia and Wallachia had only one chamber each known under the name of the General Assembly, and deputies were elected to it exclusively from the boyars, two from the landowners of each district and one from each district town. Moreover, the metropolitan presided; the eparchial bishops were also members. There are two of them in Moldavia: the bishop of Roman, who lives in the city of Roman, and the bishop of Huşi, who resides in the city of Huşi. There are three in Wallachia: the bishop of Buzău (his residence is in the city of Buzău), of Râmnic (residing in the city of Râmnic) and of Argeş in Little Wallachia, the capital city of which is Craiova. This General Assembly had legislative authority, but its resolutions, in order to have legislative force, had to be approved by the hospodar, who first had to obtain the approval of both courts. This condition, introduced into the Organic

* This Sukhtelen was the son of Count Sukhtelen, ambassador in Stockholm for many years.

Statute after its publication, was, of course, not to the liking of many Rumanians, but it was considered necessary to keep Moldo-Wallachian political independence within definite bounds.

In spite of this restriction, which, actually was insignificant—I do not recall that either we or the Porte at any time objected to the implementation of any draft law—the Danubian Principalities enjoyed complete administrative autonomy. If this self-government failed to benefit the region to the extent desired, it is because there were very few honest and moral people at that time among the Moldo-Wallachians, and the hospodars elected from them did not live up to their high calling. They combined the executive and legal authority (which then were not yet separated) in their hands and took advantage of this fact to enrich themselves quickly because they could not count on the security of their throne.

I have already spoken of the personal gain which the nimble and shrewd Prince Mihail Sturdza derived for himself from this position. Once he was in a difficult situation after one of his unscrupulous acts was revealed. I do not recall in what connection, but it seems that he took a large bribe, something like ten thousand chervonets, from the bishop of Roman, but did not fulfill his promises. The bishop raised a protest and filed a complaint with the Russian government. Our consul general, Baron Rikman (this was at the beginning of the 'thirties) was ordered to go to Jassy to investigate the case. As a result of this enquiry Baron Rikman forced the hospodar to return the money to the bishop. Everyone expected that Prince Sturdza would not remain on the throne after such a disgrace. The boyars were aroused. Preparations were made in the General Assembly to impeach him, but, as usual, the result of the debates was that the chamber presented the hospodar with a citation, which he himself prepared, thanking him heartily for his wise and beneficial rule! Even his bribe was not lost. A few years later Prince Sturdza succeeded in removing the kind and honorable metropolitan Veniamin and in rigging the election for his successor in favor of the same unscrupulous bishop of Roman; Sturdza received for this service not only the ten thousand chervonets, but twice or three times as much. Moreover, on this occasion he was handsomely rewarded by the ecclesiastical official (who was, if I recall correctly, Archimandrite Rosetti) who took the [now available] post of the bishop of Roman. These examples give an idea of the morals of the Moldavian clergy at that time.

In former days the white clergy * was even in a worse state. It was almost without education, which was probably the reason for the religious indifference among all classes of the Moldo-Walla-chian people.

Except for the most important holidays, the churches were prac-tically always empty. Even the *țărani* seldom looked into them. On the other hand, the taverns, which were usually located near the churches in the towns and villages, were always filled on holidays, even during the hours of Mass.

Nevertheless, the Rumanians, except the young people who were educated in western Europe, strictly observed the fasts; they liked the pomp of the church rituals, particularly the christenings, wed-dings, and funerals. I frequently had occasion to be present at these ceremonies during my long stay in the principalities. All those present at a christening in a prosperous family receive small gold coins or medals on pretty ribbons in memory of the happy event. The weddings are different from ours: in place of a veil the bride wears long, thickly woven gold threads (*fils d'or*), which fall from the nape of the neck almost down to the floor. Unfortunately, this beautiful, ancient custom began to change even during my stay in some aristocratic families, whose daughters, educated in Paris, preferred to wear veils. Funerals are always conducted with great pomp. This is very profitable for the clergy, who gather on such occasions in great numbers, particularly if the deceased be-longed to a distinguished and rich family. On such occasions alms are also distributed generously among the poor. I witnessed the funerals of such persons as Ion Sturdza (who was hospodar for a time before the introduction of the Organic Statute), the boyar, Alecu Ghica, and others, which cost at least twenty thousand rubles in silver. The procession begins with musicians playing a funeral march, then follow the priests in incredible numbers, sometimes coming from the most remote provinces for this occasion. They carry the top of the casket. Then follows the hearse with the de-

---

* In the Orthodox Church the black clergy are the monks; the white clergy who serve the people directly are allowed to marry.

ceased in an open casket, which makes a most unpleasant impression, particularly if the hearse rocks on a bumpy road. The mourners stand around the hearse in white clothing with their hair down. They raise a dreadful howl, each one in turn, and sometimes all together, which is interrupted by unpleasant Oriental singing by the *prichetniki*. There are no choirs either in Greek or Rumanian churches. The mourners are chosen from the house servants, and sometimes additional ones are hired. However, this old custom began to disappear during my stay in the principalities.

Speaking of funerals, I recall the death of the wife of the boyar, Costache Paşcanu (Cantacuzino), the mother of our friend Pulcheria (Pulcherie) Arghiropulo, and the wedding of the latter. These two events followed each other at an interval of two or three days. This is how it happened. Feeling the approach of death, Mme. Paşcanu suddenly expressed the wish to see her younger daughter, Pulcheria, married while Mme. Paşcanu was still alive, and only to a Greek. Her older daughter, the beautiful Eufrosina, was already married to a Phanariot Greek, Prince Callimachi. She preferred Greek men because she regarded them as more capable of conjugal happiness than the Moldavian men, who were not noted for making good husbands and who treated divorce lightly. To please the sick woman, a search was immediately made for a bridegroom among the first-ranking families in Greece, and the young Emanoil Arghiropulo was found, who, having no money at all, accepted with pleasure the proposal of a rich bride. He rushed immediately to Jassy, came at once to the home of the Paşcanus, and introduced himself to his bride, whom he saw for the first time in his life. It was decided then and there that, in view of the critical condition of the mother, the wedding should take place the same day. Since I knew nothing of the arrival of Arghiropulo and expected to hear the sad news of the death of the unfortunate sufferer, who had been in a hopeless condition for some time (I think she had dropsy), I was extremely amazed to receive an invitation to this wedding. Despite the presence of the sick woman, who was wheeled into the ballroom on a lounge chair, the wedding passed happily for the young people, who liked each other immediately. On the insistence of the dying woman, the wedding was followed by dances, and I still recall the young and attractive couple in the whirlwind of a waltz. Naturally the evening ended early, and the happily married couple were escorted to the apartment reserved for them. The funeral was three or four days later, and shortly after that the Arghiropulo couple went abroad. They used to come to Moldavia only occasionally and for brief stays while I was there, and I therefore saw them seldom. My family's closer acquaintance and even friendship with the kind and intelligent Mme. Arghiropulo began much later and under different circumstances. About her husband, I can only say that I always regarded him as an honorable and educated man. He was particularly distinguished by his talent in music and literature, and by an unusually sharp wit.

Marriages of this kind, although not with the same wild haste, were a common occurrence during my stay in Moldavia. I knew many young women whose marriage with Greek men came as a result of correspondence only. I have already mentioned the reason why they were preferred to Moldavian men. On the other hand, for the young, educated, and generally gifted Greek men, who had few opportunities in their poor country to make a good marriage, the rich Moldavian brides were a great attraction. As a rule, the Greeks finally settled there. That is why one finds so many family names of Greek origin among the Rumanian landowners.

There are also many foreigners among the people of the middle class, the merchants and the artisans. One will find here Bulgarians, Serbians, Armenians (chiefly in Moldavia), and Jews. The only pure Rumanians are the peasants. Although the Rumanians are unquestionably Romance or Latin by origin, they resemble the Slavs; and their language has many Slavic words. Their faith, social structure, character, mode of living—everything, relates the Moldo-Wallachian to the Russian and particularly to the Ukrainian. The Moldavian village differs little from a Little Russian village. The same scattered *mazanki,* the thatch roofs, the well, the household utensils, the attire, the textiles with a red motive, even the melodies of their songs are similar. Just as in Little Russia, you see white oxen with enormous horns here. A *ţăran* is just as light-hearted as our Little Russian *muzhik.* Only occasionally in some localities are the faces different, whether from the admixture of Hungarian or gypsy blood. The sight of the awkward buffalo in the Rumanian village is perhaps unusual to our eye. The milk of the buffalo is much richer than cow's milk, and the cream or *caimac* has an exceptionally fine taste.

I had just as good a time in the winter of 1843 as in the previous year. I hardly had time to reply to the invitations to parties and dinners. I have encountered nowhere such hospitality as in those times in Moldavia. By spring, the gay Cantacuzinos, from whom I was inseparable, had come from Bălţăteşti, and then the really sumptuous feasts began. In the time free from *amusements* I continued my talks with the intelligent Alexander Moruzi with whom I discussed contemporary political events. The Eastern question was the chief topic of our discussions, which were useful to me.

With the arrival of summer practically all my good friends left; some went abroad, others to the country. Kotsebu received a leave of absence and went to Courland with his family, leaving their tiny daughter Zoia in the charge of Dr. Cihac and his wife. Kotsebu asked me to visit the little child and write him about her. I gladly complied with this commission which gave me a good excuse for frequent correspondence with Kotsebu and brought me even closer to this charming family.*

The princes Cantacuzino, Grigore and Leon, decided to speculate. They leased the Holy Sepulcher estates in Moldavia and organized their own estates in Bessarabia. Naturally, nothing came of this. Grigore went to Constantinople for this purpose, lived there several weeks, spent a great deal of money, and returned with his pockets empty and with nothing accomplished. Leon's trip to Bessarabia was just as unsuccessful. Desiring to find some occupation, however (at that time the old Prince Cantacuzino had not as yet distributed the inheritance among the children), both brothers leased two private estates (*moşii*) near the city of Fălticeni: Şoldăneşti, where Grigore took up residence, and Lămăşeni, into which Leon moved. They left Bălţăteşti and the beautiful Hangu, which their father managed himself.

Feeling lonesome in the deserted city, I went to my friends, the Cantacuzinos, and spent several weeks with them in the summer.

---

* I accompanied them as far as the city of Czernowitz (in Bukovina); on the way back I stopped to see a friend, Alexander Moruzi, on his estate Zvorăneşti (near the city of Botoşani). I spent two days with him.

Although Şoldăneşti and Lămăşeni could not compare with Bălţă-teşti in their location, and even less so with Hangu, the country-side was nevertheless also lovely there. Moreover, one could not feel lonely in the cheerful company of the Cantacuzinos. Every day brought some new activity. We visited the city of Fălticeni during the fair, which is famous in Moldavia, and we had the wild-est time there.

Not far from Lămăşeni on a beautiful estate, Baia (on the river Moldova), lived Prince Dimitrie Cantacuzino (a cousin of my friends) who had returned from Bavaria. He was married to the daughter of Count Armansperg, the former regent of Greece dur-ing the minority of King Otto. Her name was Sofia. She was a strikingly handsome, and well-educated woman, but not of very strict morals. Their marriage could not be called a happy one. The brother of the owner, Prince Alexander Cantacuzino, about whom I already had occasion to speak,* also lived at Baia. We gathered frequently at Baia and spent a pleasant time. Countess Sofia was an excellent musician and often accompanied their house

---

* These Cantacuzinos were the sons of Prince Alexander Cantacuzino, the older brother of Prince Egor Matveevich. He was married, as I said, to Elisaveta Mikhailovna Daragan. The following are the names of their chil-dren: (1) Matei, married to Ernestina Ghica, the daughter of the Moldavian boyar, George Ghica. Both husband and wife had the misfortune of losing their minds almost simultaneously and are now in Germany in institutions for the insane. They have two children, a daughter Elena (Lili), married to her cousin, Alexander (the son of Prince Dmitri), an officer in the Austrian army, and a son, George. (2) Mihail, the owner of Ottaki and formerly a marshal in the Bessarabian nobility. He too was married to one of the daughters of Count Armansperg and, after her death, to the daughter of the acting governor general of the Novorossisk region, Fedorov. He had a son, Paul, by her. When he had the misfortune of also losing his second wife, he married one of the chambermaids in his employ and paid dearly for this foolishness. (3) Dmitri, married to Countess Sofia Armansperg. In ad-dition to Alexander, just mentioned, they also had two sons and a daughter, Irina. I lost track of them altogether. (4) Alexander was married to the daughter of the Moldavian boyar Nicholas Canta, about whom we have already spoken. (5) Elena, married to Count O'Donell, an officer in the Austrian army. She died from consumption in Venice in 1844 or 1845. (6) Aristid, an officer of the Russian general staff, who graduated from the Military Academy. This young man, remarkable in every respect, had the

physician, Dr. Tremel [Trömel], who played the violin. He came with her from Bavaria.

Once the Cantacuzinos and I started out on a trip into the mountains. After spending the night at Baia, we started out on horseback. Because of a great flood, it was impossible to cross the river Moldova any other way than by wading and swimming across. The ladies were, of course, not with us. We had to swim for some time. My horse was exhausted to the extent that the current began to carry it away. Realizing my danger, the two *plăiaşi* jumped in at once and, with difficulty, caught up with me and grabbed my horse by the reins. Thanks to them I reached the shore, but I was drenched. In this state I rode for several hours more, since I had no means of changing my clothes. I am surprised that I did not catch cold as a result of this adventure.

We used to undertake all these trips and engage in diversions in a large group. Our circle consisted of Prince Panaioti Moruzi (the brother of Alexander) and the brothers Canta (Cantacuzino), sons of the intelligent and cunning *logofăt*, Nicholas Canta. These young people were brought up in Paris and in Switzerland and were distinguished by their education and their worldly manners. They belonged to the new generation in Moldavia and had an entirely European polish. The wife of Iancu Canta, or Iancuşor, was the life of our crowd. Although she was far from being a beauty, she was liked because of her expressive face and lively manner. The Canta family also lived at that time not far from Fălticeni on a beautiful estate, Horodniceni. If I am not mistaken, it was in the same year, 1843, that they suffered a terrible misfortune. The entire family once were sitting as usual after dinner on the porch facing the garden where there was a small pond. Matvei Canta, the fourth brother, a young man of twenty, wanted to take a ride in a boat with his fifteen-year-old sister, Elena. After he had pulled off shore a little, he decided to scare his sister and began to rock the boat, but he was not careful, and both fell in the water. To save his sister, Matvei, who was an excellent swimmer, tried to pull her

---

misfortune of losing his mind. I can well imagine the grief of the unfortunate mother. She lived to a very old age and bore her misfortune with true Christian resignation.

after him, but in her panic she clung to him so hard that he was unable to swim, and both went down to the bottom. All this took place in a moment, in sight of the parents, the entire family, and many servants. Aid did not come in time, and they dragged out two corpses. The pond was not even one sazhen deep [six feet].

In 1843, the third son of Prince Egor Matveevich came to Moldavia, Prince Egor Egorovich, or George Cantacuzino, as everybody called him. Just as his older brother, he served in the Cuirassier Regiment of her Imperial Highness, the Grand Duchess Elena Pavlovna. He retired with the rank of lieutenant. The drudgery of service was not to the taste of my friends the Cantacuzinos. Freedom and independence attracted them more than anything else, and, with the means they had at their disposal then and which they thought they would always have, they believed that they could enjoy these blessings eternally with no cares. With their characteristic lightheartedness, they had to face, unfortunately, not only great disappointment but also the saddest consequences. However, let us not run ahead of our account, but, speaking of the joyful time for the Cantacuzinos which I am now describing, we shall say that George also proved to be just as charming and cordial a young man as his older brothers, and I became great friends with him also, just as with the rest of the family.

When I returned to Jassy, I resumed my work. During Kotsebu's absence, Tumanskii was in charge of the consulate. However, in spite of his remarkable gifts, he could not overcome his indolence, and I had to bear practically the entire load of the official duties. Tumanskii disliked even personal negotiations with the hospodar, particularly those in connection with lawsuits which called for careful investigation. I frequently had to carry on negotiations for him. Endowed by nature with great poetic talent, he had great promise . . . [defective copy]. He sought diversion in cards and sensual pleasures. Could this be expected of . . . [defective copy] the author of these magnificent lines:

> Yesterday I flung the jail door open
> To my little prisoner from the air
> To the woodlands I returned the songstress
> I returned her liberty to her.

> She vanished drowning in the
> Radiance of the blue day
> And, as she flew away,
> Into a song broke she
> As if to pray for me.

However, despite his indolence and indifference, the corpulent Tumanskii was a delightful conversationalist. Frequently, he would make intelligent and sharp observations.

Another friend of mine connected with my official duties in Moldavia, our postmaster, Alexander Khristianovich Pekgol'd, was also a unique gentleman, but in an entirely different way. Educated according to the French ideas of the last century, the kind and charming Pekgol'd was in his youth private secretary to Prince Razumovskii and was with him at the Congress of Vienna in 1815. After the death of Razumovskii, early in the eighteen thirties, he entered the service in the Postal Department as foreign postmaster in Jassy, a permanent and pleasant position under the conditions existing then. Our Alexander Khristianovich was a typical old bachelor, respectable, accurate to the point of pedantry, and honest in every respect. He made his appearance the first thing in the morning and returned home only late at night. Everybody loved him and respected his excellent qualities. He had but one fault: He was easily offended and violently quick-tempered. We in the consulate were on the friendliest terms with him, despite frequent bickering which usually was over trivial matters. The calm Tumanskii took particular pleasure in enraging him. I recall once an argument between Pekgol'd and Tumanskii as to who was the better connoisseur of wines. "Never," said Pekgol'd, "could I make a mistake in the quality of a wine, and I can always tell its vintage by tasting it. I am ready to bet on this. Nobody can fool me." Tumanskii accepted the wager, the loser to buy a bottle of champagne. Some time passed and the matter seemed to have been forgotten, but one day when all of us were to have dinner at the hotel, Tumanskii took advantage of the occasion. He went there in the morning and arranged to have the best *Cotnar* (Moldavian) wine poured into an empty bottle of Haute-Sauterne. The bottle, label and all, were then artfully resealed so that there was no obvious evidence of the fact that it had been used before. Suspecting nothing, our poor Pekgosha (as we called him in our intimate circle) was completely fooled. Our teasing when we succeeded in the joke enraged him beyond words. He was so furious that he rose from the table, swore at all of us like a trooper, and left. Similar scenes were frequently repeated. However, the anger of the kind Pekgol'd passed quickly, and we always remained good friends.

My third comrade, Evgraf Romanovich Shchulepnikov, adapted himself so well to the mode of life, customs, and manners of the region, which were much to his liking, that he soon became completely Moldavian. He was in touch with all classes of society, which may be the reason for my seeing him in the upper circles rather seldom. He led a dissipated life, and it is a pity that he did not make the most of his excellent gifts. He was in every sense of the word a fine young man and we had the pleasantest relations in spite of some faults, which must be attributed to his flightiness and irresponsibility. Seldom did I have occasion to laugh as much as I did with him. He was thinking of settling down and having children, since he thought he had the qualities of a good husband. Marriage was his favorite dream, which, considering his mode of life, seemed funny. Incidentally, he liked a sweet young lady, Emilia, the daughter of Dr. Cihac, and since I saw her almost daily when visiting her parents, who were taking care of Kotsebu's little daughter, Shchulepnikov asked me to say a good word for him to Emilia. I willingly undertook the errand to do a good turn for my friend. At the first opportunity, when we were alone, I began to tell the young lady that she inspired a flaming passion in a young man, a close friend of mine, who would consider himself very happy if she would cast her lot with his. Apparently I was successful, because Mlle. Emilia, thinking that I had myself in mind, showed signs of accepting the offer. When I noticed this, I hastened to explain and named Shchulepnikov. She changed her tone completely and declared that she would never hear of it. It was hard for me to tell Shchulepnikov about my failure in the delicate talk with the young lady, the results of which he awaited with great impatience. However, when I told him everything in detail, he was so amused by the misunderstanding (*quid pro quo*) that he forgot the main object of the talk. So was our flippant Evgraf Romanovich all his life.

Speaking of the summer of 1843 brings to my mind the great fire which destroyed a large part of the city of Jassy. It started on Main Street, came close to our consul's home, and spread over the entire outskirts behind the St. Spiridon church. Several hundred houses burned down, most of them with shingle roofs. After this fire the Moldavians began to build more sturdy houses and with iron roofs almost throughout, which up to that time were quite rare in Jassy. We were merely frightened, but the Prussian consul general, Neigebauer, who was absent at the time, lost practically all possessions. With the aid of our consular cossacks, I succeeded in saving his archives, correspondence, and everything that seemed to me to be the more valuable of his possessions. Upon his return to Jassy Neigebauer expressed his hearty appreciation to me and recommended me for the Prussian order of the Red Eagle, but for some reason I never received it. Apparently the Berlin cabinet did not attribute too much importance to my rescue of the archives.

The end of the summer before the return of the Kotsebu family was a lonely time for me. Occasionally I went to the hospitable elderly lady Bogdan and her daughter, Maria Balş, who also entertained often.* Her husband, Hetman Todoraş Balş, at one time commanded the militia under General Kiselev. He was one of the first men in Moldavia to shave off his beard and to wear European clothing. He liked to give himself the air of an old soldier and always wore spurs and an army cap. He was never businesslike and, consequently, despite his high position, never played any political role in the region. He was one of the most sincere supporters of Russia.

On my trips to Skuliany I dropped in occasionally to see Nicholas Rosnovanu in his magnificent castle, Stânca, which I had occasion to mention earlier. However, I did not go often to Skuliany because, for the most part, I was busy with my secretarial duties at the con-

---

* Smărăgdiţa Bogdan also had two sons: Manolache, kind but weakminded, and Lascar, handsome but not too bright. He [the father] served in his youth in the militia, but in my time he was a retired colonel.

sulate. I used to send our civilian functionary, Komarnitskii, to be present in my place during the opening of our dispatches from Jassy, Bucharest and, particularly, those from Constantinople, in the quarantine. I could depend entirely on the discretion of this official.

By the end of September the Kotsebu family had returned from Courland. Kotsebu himself came back a few weeks later because he decided to travel a little through Germany. Because of this circumstance, Leon Cantacuzino's wife, Emilia, hurried to Jassy to be with her sister and remained with her for some time. I was like a member of the family there, always in the pleasant company of these two ladies. Taking advantage of the beautiful fall days, we took walks daily in the countryside surrounding the city.

The winter of 1843–44 found us indulging in the same recreations which I shall not attempt to describe because it would be repetitious. One could not expect too much variety in a small town like Jassy. The only new event was that Iakov Andreevich Dashkov visited us that winter, and there were many parties on that occasion. Conditions have changed, and the Russian representatives do not enjoy the same respect that they did then.

About this time the hospodar's sons returned to Jassy from their education abroad—the *beizadele* Dimitrie and Grigore Sturdza. The former proved to be a kind and harmless young man. Since his inclinations were toward military service and since he spent his days at the military stables training his riding horses according to the Boshe system, he was appointed hatman, or commander in chief of the Moldavian militia. As for *Beizadea* Grigore, who possessed a remarkably fine mind and a strong will, he instilled fear in everybody, including his domineering parent, who was at a loss to know how to manage him. His vanity knew no bounds. He barely had time to receive the appointment of his choice in one of the ministries than he announced that he was dissatisfied with the maintenance assigned him by his father. He then decided to enrich himself quickly by his own means. He conceived the idea of leasing church lands, even employing coercive measures to this end. Traveling through the monasteries, he forced the owners to lease the choicest lands to him and on the most favorable conditions, which he fulfilled only if he chose to do so. Surrounded by a few armed arnăuts, he ruled like a feudal lord, a law unto himself, paying no attention either to the police or the judges, all of whom trembled before him. Few complaints ever reached the hospodar, but even he did not dare take any steps against Grigore. This lasted for some time until we succeeded in moderating his actions through the influence of our consulate. With the years the *Beizadea* Grigore changed for the better. I shall have occasion more than once to speak about this unusual person in these notes.

About this time also the brother of Countess Sturdza, Nicholas Vogorides, came from Constantinople, the older son of the Prince of Samos (Island of Samos). He came with the intention of improving his finances through a good marriage, and, thanks to his connections, married the only daughter of one of the richest boyars in Moldavia, Costache Conachi. Prince Nicholas Vogorides, a fine looking man, was a typical dandy of Constantinople. He wore his fez (a red Turkish cap) on the side and from under it his carefully waved black hair fell down to his shoulders. He wore his half-Oriental clothes with great elegance. He was a kind but empty man and an extreme squanderer. In this respect he presented a glaring contrast to his father-in-law who amassed a fortune through his thriftiness. Because of his stinginess, the old Conachi even made his own little caps, which the Moldavians wear at home as part of their national costume.

Speaking of the Moldavian national costume, no one wore it as well as the *vornic* George Balş, known by his nickname, Balşuţ. I met him almost daily at the home of Kotsebu, with whom he always played whist. Balşuţ was a great eccentric. Having dissipated a large fortune, he was forced to live poorly and deprive himself of much. But the one thing which this old bachelor refused to deny himself was elegant dress. His clothes were made of the most costly materials, and he had a great number of magnificent cashmere shaws which he used as sashes. The elderly Balşuţ—he was then around seventy—was well educated for his time. He knew the classics by heart, not only the ancient ones in the original, but also the French. He was of a cheerful disposition and enjoyed telling stories of the old times—he did witness a thing or two in his youth! He knew personally our famous generals—Kamenskii, Miloradovich, Kutuzov, Suvorov, and even Potemkin.* I learned a great deal about Suvorov also from Costache Conachi who entertained him while he was district police officer in Tecuci. "Once," Conachi said, "I received through the scouts the information that the Turks were marching in large formations from the direction of Focşani and were no more than a distance of a few versts from us. It happened to be a holiday and Suvorov was in church. I ran to him and passed on the terrible news to him. Entirely unperturbed, Suvorov, who sat in the choir as usual, continued to sing as if nothing had happened. Only after the services were over did he call me and ask me to tell him everything in detail. He then calmly gave orders to his men to prepare for a counterattack against the enemy and, although completely unprepared, gained a decisive victory over the numerically larger Turkish army." This account by an eyewitness, who had personal dealings with Suvorov on the eve of the glorious battles near Focşani and Râmnic is without a doubt of great interest.

Among the evening visitors at the Kotsebus who spent their time at cards, I forgot to mention the aged Gaetan Zelinskii, almost one hundred years old. All his life he had been an inveterate, passionate player, spending whole nights over cards. However, in spite of such a physically and morally dissipating existence, he was still vigorous and tireless! He always kept the bank and not one card escaped his eagle eye, although there were a great number of punters. He came to Moldavia in 1828, if I am not mistaken, with Senator Krasno-Miloshevich as an official on special mission. The boyars liked the game for its own sake. They were quite a source of profit to Zelinskii who retired and settled down in Moldavia. However, everyone regarded him as an honest player and a kind man. He had known my father well during his stay in Volynia and he told me many things about him. Zelinskii endeared himself to me because of his excellent opinion of my unforgettable parents. He died in Jassy in 1847, if I remember correctly.

The name of Zelinskii recalls to my memory one unpleasant circumstance which caused me to lose a lot of money. There was a Pole at the French consulate employed as dragoman whose name

* How many interesting anecdotes did he not tell me about them!

was the same but who was not related to the old card player. This young Zelinskii had a bad reputation as a bribe taker and a usurer. He loaned money at a fabulous rate of interest. I did not have the pleasure of knowing him, nor did I ever come to him for money. But I did have the misfortune of dealing with him unexpectedly, and this is how.

There lived in Jassy one Count Gabriel de St. André, who belonged to one of the best French families of Saint-Germain. He posed as a legitimist emigrant who had left his native country as a consequence of the revolution of 1830, appeared to be a respectable gentleman and, consequently, was received in the best homes. The count did not have much money because he had dissipated and lost his fortune in his youth, and merely received an insignificant pension from his relatives in France. He came to Zelinskii one day and asked to borrow two hundred fifty chervonets. The latter agreed, but on the condition that St. André present a responsible cosigner. Tumanskii, with whom our *émigré* was on friendly terms, was unable to help him because he himself owed money to Zelinskii. Tumanskii suggested my name. Because of my lack of experience in such matters, I agreed to vouch for the man. I did this, however, mostly because of Tumanskii's assurances that St. André's honesty and promptness were beyond doubt. However, no more than two months after I had given my signature St. André was arrested and surrendered to France. He succeeded in escaping on the way. I do not know where he vanished, but he never returned to Moldavia. The fact was that on the accusation of his wife, from whom he had been separated for some time, [the French] had condemned him to hard labor for forging her signature on checks and other documents. As a consequence, the French government requested his extradition. Since St. André did not pay the debt to Zelinskii, the responsibility fell upon me. This disrupted my finances for a long time, because with my modest salary it was hard for me to pay two hundred fifty chervonets, and interest besides. To make matters worse, in addition to the painful loss of the money, I was in danger of jeopardizing my reputation with my principal superiors because the kind Karl Evstaf'evich Kotsebu, thinking it unfair that I should have to pay for St. André, expressed himself to that effect to the French consulate in reply to their request for settlement of the debt. The matter was referred to Constantinople, and our ambassador condemned me. The rumor spread in our embassy that it was I who had borrowed the money to pay a gambling debt. All of this was extremely unpleasant.

Our intimate circle was widened at that time with the arrival of a young German scholar, Foveger, brought from Breslau to tutor Kotsebu's sons. My conversations with this remarkably intelligent young man were of great interest to me. Although I did not always share the radical ideas he held at that time, I found great satisfaction in listening to this enthusiastic German. I was quite opposed to his convictions and constantly argued with him. This, however, in no way affected our friendly relations. The profound scholarship of Foveger, a candidate for a Doctor of Philosophy degree, did not interfere with his cheerful disposition. One was never bored in his company. Such a friend was a real treasure in a land where genuinely enlightened people from whom one could learn were seldom encountered.

I should also like to mention Tavel, a Swiss, with whom I became good friends. He taught French to Kotsebu's children and was a fairly educated man, but carefree and lazy. As I recall, my friends the Cantacuzinos sent him in 1844 to Ottaki in Bessarabia to teach French to their sister, the young princess Olga Egorovna, and her friend, Princess Sofia Krasnakutskii, who was the niece of Princess Elisaveta Mikhailovna Cantacuzino.

In the spring of 1844 I went to Khotin for an interview with Kolonov about the following matter.

As I have mentioned, my brother-in-law, Captain Kolonov, commanded a battalion of the Home Guards in Kamenets. The tsar was expected there in the summer or early in the fall of 1842. Needless to say, as an old experienced army man, Kolonov took all the steps necessary to present his army unit in the most brilliant manner during the tsar's review. Unfortunately, he fell gravely ill just before the arrival of His Imperial Majesty and could not appear for the review. The old staff officer who replaced him lined the soldiers up for review in brand-new uniforms, leaving the new recruits, who had just arrived in the barracks, in the oldest, most worn-out coats. He thought that he would have enough time to bring the battalion in from review and form the guard in the proper manner. To his misfortune, but particularly to the misfortune of Kolonov, the tsar decided to stop at the barracks first before going to the review. What he saw there would have shocked anyone: sentries in rags and not even in complete uniforms. Moreover, not one of the men knew how to salute properly. The tsar's wrath was beyond description, and, refusing to accept any explanation, he ordered Kolonov's immediate arrest. Hardly had Nicholas Ivanovich had time to recover from his illness than he was incarcerated in the Khotin fortress. My poor sister, Iulia, therefore, came to live in Khotin. The results of a thorough investigation revealed that Kolonov's battalion could have served as the model for the entire army and that the person responsible for the affair which caused the tsar's righteous indignation, was the staff officer who temporarily had replaced Kolonov. The case dragged on for almost two years. Nicholas Ivanovich was not only acquitted, but even praised. He was offered a choice of services or a civil post commensurate with his rank, but he felt so disgraced that he insisted on a straight retirement. When I decided to visit the poor Kolonovs and share with them their grief, the case was drawing to a close and Nicholas Ivanovich was permitted to live with his family in a private home, without, however, leaving the city of Khotin.

I went by way of Skuliany, where I had to spend four days in quarantine. I was given two rooms for myself and my man (I think it was the Moldavian who replaced Maksim). A guard, *gvardian*, was assigned to us. We were not permitted to communicate with anyone. The food we received from the canteen keeper was passed to us with all due precautions. The kind Alexander Grigor'evich Rozhalin, the inspector of the quarantine, visited me daily. We talked, however, in the courtyard at a respectable distance from each other. It is hard to imagine anything more tedious and agonizing than this isolation. I spent almost the entire day reading. On leaving the quarantine I went to Khotin by the relay horses.

The part of Bessarabia that I crossed differs from Moldavia only in that the post-horse transportation and the stations are similar to the Russian style. The population is the same—Rumanian with only some admixture of Little Russian. Gypsy camps were everywhere along the way. The Russian language is prevalent in the cities, but the natives, or *țărani*, seldom understood Russian. The villages look exactly like those in Moldavia—scattered *mazanki,* and practically every one of them with a stork's nest. This long-legged bird, with its enormous beak, is encountered at every step. It was early spring when I made this trip, and the storks had just returned from the warm countries where they migrate for the winter. It is astonishing that after an absence lasting almost half a year each stork invariably finds its own nest, and the *țăran* receives his old friend with pleasure.

After visiting the Kolonovs for about ten days, I returned to Moldavia. However, this time not by relay horse, which was a tiring method that exhausted me. Instead I took a long ride on a Jewish wagon in which I was dragged for days over narrow roads to the little frontier town of Lipcani on the Prut, from where I proceeded through the Moldavian district town of Huşi to Jassy. Traveling in such a wagon is very peaceful, but it was tiresome because the Jew often stopped to give the horses a drink or a rest, and I could not cover more than forty versts a day.

I carried away a very melancholy impression of this journey. It is more than thirty years since we took over Bessarabia, and what have we done to contribute toward its welfare? Have we introduced new laws in the administration, in the courts? Have we done anything to civilize the region? Have we built new roads? Alas, to all these questions we must give a most emphatic negative answer. My patriotic feelings suffered greatly when I compared Bessarabia with the other province also torn away from Moldavia—Bukovina, where the Austrians succeeded in introducing such exemplary order in every respect!

I kept in touch with my relatives, particularly with my brother Alexander whom I wrote at least once a week from the time of my arrival in Moldavia. My letters to him for this period constituted something of a journal. I related in detail to him my observations and impressions, without leaving out any event which might have been of some interest. I regret that our correspondence has not been preserved. It would have greatly facilitated my present work. Shortly after our separation my brother Alexander was transferred to the Ministry of Interior. In his capacity as an official on special commissions attached to the ministry, he traveled much through the provinces either on tours of inspection or to form new institutions. He made an excellent impression in his post and soon became one of the best functionaries of his department. He was as important as Nicholas Alexeevich Miliutin, Iakov Khanykov, N. I. Nadezhdin, and others like them both in regard to their talents and to their lofty and liberal tendencies which they followed under the guidance of the minister of interior, Count Lev Perovski.

My brother Fedor was then (in 1844) completing his studies in the School of Jurisprudence, where he was regarded as one of the best students.

Of my other relatives in Petersburg, my uncles, Ivan Savvich Sul'menev and F. P. Litke, wrote me occasionally.

At that time (1841–1845) everything was going well in the household of the Sul'menevs—thank God! However, a great misfortune befell Uncle Fedor Petrovich. He lost his worthy, charming, and beloved wife, Iulia Vasil'evna. She passed away in Tsarskoe Selo on September 8, 1843, exactly at the minute when a cannonade announced to the people the birth of the Grand Duke (subsequently heir and tsarevich) Nicholas Alexandrovich, this excellent youth, so promising, who died prematurely in 1865 in Nice. He was, as is well known, the happy fiancé of the Danish princess Dagmar (Maria Fedorovna) [who after his death became] the wife of the present heir [the future Alexander III] . Aunt Natalia Petrovna Sul'menev was at the death bed of Iulia Vasil'evna. She took the small children, Konstantin and Nicholas Litke, with her to Petersburg.

In 1842 my dear uncle, Fedor Karlovich Giers, also died. I preserved a letter that I received from him on June 10 of the same year in which he talked of his great suffering. He had a painful disease, gall stones. After his death, his only daughter, Valeria, was taken to Petersburg by our aunt, Anna Karlovna Maiet. They settled there with the two sons of Anna Karlovna, Fedor and Konstantin Petrovich. The latter married Valeria several years later. She was his cousin.

Nothing in particular happened to my other relatives during that time. The Speranskiis continued to live in Berdichev. The Anferovs moved to Poltava where Nikifor Ivanovich was appointed prosecutor. I have already mentioned the Kolonovs, and my youngest sister, Anneta, was still at the Catherine Institute.

Let us now return to Moldavia where I was beginning to feel like a long resident.

My life there continued as usual, but the diversions of society began to bore me and I spent most of my time in the family circle of the Kotsebus and the princes Cantacuzino whom I visited often in the country.

The Kotsebus spent the summer of 1844 again in Socola in the dacha owned by the metropolitan. I used to visit them daily.

During this summer Prince Egor Matveevich Cantacuzino went to live in Germany and left all his estates under the management of his sons Grigore and Leon, also his son-in-law Vasilii Evstaf'evich (Guillaume) Kotsebu, married to his older daughter, Princess Aspazia.* In accordance with the deed of cession, they were to be responsible for the care and improvement of the estates as well as

---

* Their marriage took place in 1839 under the following circumstances. Vasilii Kotsebu, a diplomatic official with the governor general of Kiev, Bibikov, came to visit his brother, Karl Evstaf'evich in Jassy. There he met Molly Kotsebu's sister, Emilia Koskul'. He fell in love with her and proposed marriage to her. However, he had a dangerous rival, Prince Leon Cantacuzino, whom Emilia preferred. Vasilii Kotsebu, grieved by the refusal, started back for Kiev. His journey was to take him through Ottaki. Near this estate one of the wheels of his coach broke down. This accident forced him to interrupt his trip. Knowing that Princess Elena Mikhailovna Cantacuzino was living there with some of the members of her family, he went to see her and was received in a most friendly manner. He met Princess Aspazia there, was captivated by her unusual beauty, and became so enchanted by her that he would not leave Ottaki. The carriage was repaired but he abandoned all thought of continuing his trip. Finally he decided to speak to Aspazia and asked for her hand. The matter was settled quickly and their wedding took place almost at the same time as that of the former object of his love Emilia Koskul' with Prince Leon Cantacuzino. From Jassy Vasilii Evstaf'evich (Guillaume) Kotsebu left for Kiev in the company of his wife. About a year and a half later he was appointed secretary of the mission in Karlsruhe. Desiring to participate in the management of the Cantacuzino's extensive estates, in accordance with the surrender deed of 1844, mentioned above, Guillaume retired and took up residence in Bălțătești. This participation was of no use either to him or to the Cantacuzino family. He could not get along with his brothers-in-law. Realizing that neither harmony nor understanding could be reached in this joint management, Guillaume withdrew from it altogether. The whole business remained in the hands of Leon, whom no one dared to contradict.

the payment of the debts, which at that time had reached almost twenty thousand chervonets. They were also to pay each member of the family a definite income. The old prince reserved three thousand chervonets a year for himself and assigned a thousand chervonets each to his wife and all the children. The shares of the younger children, not as yet of age (Mihail, Nicholas, and Olga) were to be put aside to accumulate in a special fund which each of them was to receive, after deducting the cost of their education, when they became of age.* All this seemed to be well thought through and to provide fully for the future of the family. Unfortunately, in actual practice it proved to be different. It is my firm belief that the deed of 1844 was the cause of the misfortunes of the Cantacuzino family. However, more about this when I come to later circumstances in my narrative.[4] For the time being, however, let us bid farewell for a few years to the old Prince Egor Matveevich, who, arriving in Germany bought a house in Baden-Baden and took up residence there. Proximity to the roulette wheel had a bad effect on his finances.

Before his departure abroad, Prince Egor Matveevich had an interview with his wife, who lived in Bessarabia, in Skuliany. The princess arrived there with her fourteen-year-old daughter, Olga; the entire family, except Mihail and Nicholas, who were in school in Derpt, remained together for two days, I believe. I recall that at that time a portrait was painted of the young princess Olga by the artist Kaufman, whom her father brought with him to Skuliany. Through this excellently executed portrait I came to know for the first time the person with whom my fate was to be sealed forever. At the time I did not yet consider it, although I did have some premonition even then. My personal acquaintance with Olga Egorovna followed only after we met in 1846.

After the departure of Prince Matveevich to Germany, I went to the country with the Cantacuzinos and Kotsebus. I also spent several weeks in the home of Leon in Lămăşeni where the Kotsebus also stayed. We used to go almost daily to Grigore Cantacuzino's at Şoldăneşti and to Baia. All these estates are close to each other.

---

* Deducting these expenses, there should have remained a clear income from the estates amounting to several thousand chervonets.

Staying then at the home of Grigore Cantacuzino, or Grisha, as we used to call him, was his mother-in-law, the kind and jolly Ekaterina Khristoforovna Krupenskii, *née* Comnen, who arrived from Bessarabia with her two daughters, Elisaveta and Sofia. The first soon married a Greek, the *Beizadea* Dimitrachi Sutzo. A year or two later the second daughter also married a Greek, Alcibiade Arghiropulo. If I am not mistaken, that same year Grisha was visited by the brothers of his wife, Nicholas and Georgii Matveevich Krupenskii, with whom I became very friendly. Nicholas Matveevich served in the same Cuirassier Regiment as had Grigore and George Cantacuzino. Like them, he too was a lieutenant in the reserve.

I do not recall anything worth mentioning about my life in Moldavia in the period between 1844 and the summer of 1845. My duties continued as usual and, receiving neither a promotion nor the slightest encouragement from Petersburg, I began to lose hope for a decent career in the diplomatic field. All my colleagues outdistanced me, and I remained in oblivion despite the promises given me by Seniavin when I was sent to Moldavia. Solicitations in my behalf by Dashkov and Kotsebu brought no results. In 1842 I received the civil rank of *titular counselor* for my length of service, and only in October of 1844 was my appointment as agent to the Skuliany quarantine, the post for which I came to Moldavia, confirmed. This gave me an increase of four hundred rubles in salary. However, it still amounted to only slightly over more than twelve hundred rubles in silver, and that after six years of active service!

In order to extricate myself from this unpleasant situation, I began to think seriously about returning to Petersburg. However, I was unable to carry out this intention soon because Tumanskii was on an extensive leave, and I was instructed to carry his load of duties, which, incidentally, I did during his presence as well.

F. A. Tumanskii had already been in Petersburg more than a year awaiting a new appointment. He was offered the post of consul in Syra in the Archipelago and was about to go there when he was suddenly overcome with a longing for Moldavia. He declined to take the proffered post and returned to the same position in

Jassy.* I then began negotiations for a leave of absence, but received it only in January of 1847. Lev Grigor'evich Seniavin chose
to think that my presence in Jassy was indispensable until then.

While awaiting this desired leave, I languished in Moldavia.
I found my only consolation in the amiable company of Kotsebu
and my friends the Cantacuzinos. My former ardor for worldly
diversions had cooled. Thus the winter of 1844–45 passed quietly
for me. Also no doubt the fact that I spent several weeks in bed
with chicken pox must have had something to do with this. My
illness passed, thank God, without any complications, and leaving
no marks.

So far as I can recall, it was approximately around this time (no
doubt in the fall of 1844) that the Sultan Abdul Medjid made a
trip to the Danube. The hospodars Sturdza and Bibescu were
ordered to appear before him in Rushchuk to pay their respects.
They had to dress in Turkish garments and, naturally, as they are
being worn now—in *cazace* and red fezes. The magnificent full-
dress coats in the European style had to be put away. This was
quite a blow to the hospodars' vanity.

Tsar Nicholas Pavlovich sent Adjutant General Grabbe (later
count) to Rushchuk to greet the sultan. Stopping in Jassy on his

* I have preserved a letter from Tumanskii from Petersburg, dated March
15, 1845, from which one could deduce that he was yearning to return to
Moldavia. Here is what he wrote on the occasion of meeting the interesting
Moldavian Costache Prunaru: "Yesterday when taking a walk on the Nevskii
Prospect I met Costache Prunaru. You can well imagine my surprise because I was completely ignorant of the fact that he was in Petersburg. I was
charmed to see *this compatriot* of my adopted land; he came to see me today. There is an incomprehensible charm about life in Jassy, in that country, insipid though it seems, when we are there, but which we miss after
leaving it. This is a sentiment you will not fail to experience and you will not
think then that I am the only one who feels it. All those who were once
in Wallachia and Moldavia share with me this nostalgia for the principalities." Fedor Antonovich knew the principalities even before 1829 when he
served in the office of General Kiselev. He returned with him to Russia, but
in 1840 he went back there as secretary in the consulate in Moldavia. Life
in that land suited his temperament perfectly. Certainly, everyone there
learned to love him for his honorable and placid character and for his good
humor.

way, Grabbe visited K. E. Kotsebu frequently. I recall that once at the dinner table he began to scrutinize my face and finally asked me if I was related to Admiral Litke. On my affirmative reply, he said he had asked this question because he was struck by my resemblance to Fedor Petrovich. This, I must confess, flattered me.*

In the spring of 1845 Grisha and Leon Cantacuzino leased their estates of Lămăşeni and ¡Şoldăneşti and moved back to their magnificent and charming family estate, Bălţăteşti. I hastened to visit them there. They had entrusted their young brother, George, with the care of Hangu, where he had moved and where he had grown wild, leading a life of solitude. He cut an original figure with his long hair which fell down to his shoulders and which went well with his cleverly designed fantastic costume. His whole mode of life was in keeping with the primitive scenery around him. His servants were Hanguans, dressed in their mountain costumes. Bear cubs roamed freely in his yard and even wandered into the house. But our new manager was more preoccupied with hunting wild animals than with the administration of the thirty-five-thousand-desiatina estate, which was at best more than he could handle.

Early in July (1845) I learned that the Grand Duke Konstantin Nikolaevich was about to undertake a journey East and was expected to arrive in Constantinople around the 15th of this month. Knowing that the grand duke was not yet of age and could travel only accompanied and chaperoned by my uncle Fedor Petrovich Litke, I decided to join them in Constantinople. But what could I do in order to succeed in this venture in time? It was too late

---

* General Grabbe was married to a Moldavian, the daughter of Dr. Evstafi Rolla. Before that, she was married to an unimportant and plain-looking boyar, named Carp, when Grabbe met her. It was during the Turkish campaign of 1829 when Grabbe commanded the Cavalry Brigade. The beautiful Mme. Carp appealed to him. When he learned that the feeling was mutual, he decided to marry her at any cost. However, Carp would not hear of a divorce. Unperturbed by this circumstance, our ardent general took a platoon of his uhlans and arrived with them in bright daylight in front of Carp's house, carried off his wife, and drove with her directly to the church of St. Spiridon where a priest married them. The protests of poor Carp were of no avail.

to apply to the ministry. Moreover, feeling Seniavin's unkind attitude toward me, I was assured in advance of a refusal, particularly since my request could not have the support of Fedor Petrovich who was already on the way with the grand duke. Nothing remained but to write to our ambassador in Constantinople and ask permission to come there for an interview with my uncle. I did this, and in order not to lose time and not to miss the boat, I set out for Galatz, asking Ambassador Titov to address his reply to me there.

I started out in the greatest haste in a relay carriage to Galatz and arrived there shaken up and exhausted from driving over a rough road, and from the unbearable heat and dust. I stayed at the home of my good friend Sokolov, who was glad to see me. On the following day the boat from Constantinople arrived and brought Titov's letter to me. I opened the envelope with impatience, anticipating the happiness of making a journey which I had dreamed about for such a long time and which I would now take under such pleasant auspices. I also looked forward to a reunion with my dearly beloved uncle after our long separation. But alas! The letter contained an emphatic refusal. Vladimir Pavlovich Titov notified me that the grand duke had not arrived as yet in Constantinople and that, not having yet had the pleasure of seeing Admiral Litke, he did not think it possible to give me permission to come there. I cannot express the extent to which this refusal grieved and offended me. After spending two days with Sokolov who shared my sorrow, I returned shamefacedly to Jassy.

I was beginning to recover from this unpleasantness when suddenly I received the most friendly and congenial letter from Uncle Fedor Petrovich from Constantinople with an invitation to hasten as fast as possible in order to reach him there. Simultaneously Titov wrote the same thing to K. E. Kotsebu. My preparations did not take long, and the following day saw me again on the dusty Galatz road.

VIII

TRIP TO CONSTANTINOPLE (1845–1847)

IN GALATZ I FOUND the Austrian steamship "Metternich," Lloyd
Line, sailing for Constantinople. This was the ship on which I
was to make my first ocean voyage. It was filled with passengers,
and I got with difficulty a place in a cabin for four. I had to use
the upper berth which proved to be most inconvenient. However,
no difficulties could dampen the high spirits I was in as I boarded
the ship. The thought that my cherished dream to take a trip
East was being realized, and under such unusual circumstances at
that, excited me no end. The jolly company in which I found
myself also contributed to this in no small measure. Among my
fellow travelers I recall two Frenchmen—Marquis or Count Beau-
fremmont and Viscount Nittencourt, two charming young men, just
out of school. They were undertaking a journey East accompanied
by a very intelligent and educated teacher whose name I do not
recall. We became friends immediately. Seldom have I met such
pleasant and merry traveling companions. We also had a great
many Germans and other nationalities on board. On the whole
our passengers, including many tourists, were gay and interesting.

The estuary of the Danube along which we sailed divides into
three arms, the Kilia, the Sulina, and the St. George. All of them
were then in our possession. We directed our course along the
Sulina arm, the only navigable one at that time, but with great
difficulty because of the shallow water. Like the estuary of the
Nile, which forms a delta, the estuary of the Danube has sand bars,
so navigation over it was often impossible. It was Russia's responsibility to clean the estuary, but we did this for the sake of ap-
pearances only, because it was not to our advantage to make this
route easier for foreign trade with the Black Sea region to the
detriment of Odessa, whose development was rapidly proceeding
at that time. I recall that at the beginning of my service in the
Asiatic Department complaints from foreign powers with respect
to this became so insistent that in order to pacify them the Ministry

of Foreign Affairs decided to send to Sulina Active State Counselor Radofinikin (the son of the famous K. K. Radofinikin who directed our Eastern policy for many years) to investigate the question on the spot. This pacified the foreign governments, but not for long, because they soon were convinced that the Danubian commission headed by Radofinikin would achieve nothing. The question was important particularly to Austria. It is not surprising, therefore, that following the unfortunate Crimean campaign Austria succeeded in setting up in 1856 a European commission which is also active today in the estuaries of the Danube. The work of this commission, which costs tens of millions, was crowned with success, and navigation in this region no longer presents difficulties.

[Unfinished]

Notes by the Editors

PREFACE

[1] Comte C. de Nesselrode, *Lettres et papiers* . . . , 1760–1850 (Paris: A. Lahure, 1904–1912), 11 vols.

[2] These include the books, Barbara Jelavich, *Russia and the Rumanian National Cause, 1858–1859* (Bloomington: Indiana University, 1959); Charles and Barbara Jelavich, eds., *Russia in the East, 1876–1880: The Russo-Turkish War and the Kuldja Crisis as Seen Through the Letters of A. G. Jomini to N. K. Giers* (Leiden: E. J. Brill, 1959); Charles Jelavich, *Tsarist Russia and Balkan Nationalism: Russian Influence in the Internal Affairs of Serbia and Bulgaria, 1879–1886* (Berkeley and Los Angeles: University of California Press, 1958); and the articles by Charles and Barbara Jelavich, "Bismarck's Proposal for the Revival of the Dreikaiserbund in October, 1878," *Journal of Modern History*, XXIX (June, 1957), 99–101; "Jomini and the Revival of the Dreikaiserbund, 1879–1880," *Slavonic and East European Review*, XXX (June, 1957), 523–550; "Russia and Bulgaria, 1879: the Letters of A. P. Davydov to N. K. Giers," *Südost-Forschungen*, XV (1956), 427–458.

EDITORS' INTRODUCTION TO PART ONE

[1] The linguistic background of Giers was varied. As he reports in his memoirs, he and his brothers and sisters were cared for by Polish-speaking servants. His father, who had been educated in Riga, spoke and wrote German readily and used it in his correspondence with his son. His mother, in contrast, preferred French or Russian. Giers himself used French or Russian interchangeably. His memoirs are in Russian, but his voluminous correspondence with his wife is in French. His style in French was far more lucid and expressive than the writing in these memoirs which is dry and repetitive throughout.

[2] For the relationship of Nesselrode with Nicholas I, see in particular Nicholas V. Riasanovsky, *Nicholas I and Official Nationality in Russia, 1825–1855* (Berkeley and Los Angeles: University of California Press, 1959), pp. 44–46.

[3] See Barbara Jelavich, *Russia and the Rumanian National Cause, 1858–1859* (Bloomington: Indiana University, 1959), chap. 2.

I

[1] General Tormazov was in command of the Third Army or Army of Reserve which defended Russia's left or southern flank in the campaign of 1812. During the invasion of Russia, Austria was only a reluctant ally of the French. When the Napoleonic armies were forced to retreat, Austria

changed to the Russian side and joined in the coalition of powers against France.

[2] The titles in these memoirs, such as here collegiate counselor, stem from the time of Peter the Great, who in 1722 drew up a "table of ranks" for those in the civil, army, and naval services as well as for those attached to the imperial court. This reform provided for fourteen parallel civilian, army, and naval grades. In the civil service the lowest rank was collegiate recorder, or registrar, and the highest was chancelor. A collegiate counselor was sixth in rank and corresponded to a colonel in the army or a captain of the first rank in the navy. Although some additions and modifications were made in later years, the "table of ranks" was used until 1917.

For a list of the ranks and a brief discussion of their history, see Brokgauz' i Efron, "Tabel o rangakh'," *Entsiklopedicheskii Slovar* (1901), XXXII, 439–441 and "Tabel o rangakh'," *Bol'shaia Sovetskaia Entsiklopediia* (2d ed., 1956), XLI, 446–447.

[3] Prince Adam Czartoryski was a member of the highest Polish nobility who became Russian foreign minister in 1804. He exerted a strong influence on Alexander I to create an autonomous kingdom of Poland after the Napoleonic wars and attempted where possible to further Polish national interests.

[4] Count Pozzo di Borgo in October, 1814, wrote a memorandum to Alexander I against the creation of a constitutional kingdom of Poland, arguing that "the destruction of Poland as a political power shapes almost the whole of modern Russian history." Constantin de Grunwald, *Tsar Nicholas I* (New York: Macmillan, 1955), p. 99.

[5] Despite the advantages gained by the Poles at the Congress of Vienna, they continued to seek complete freedom from Russian control rather than administrative autonomy. Polish agitation finally culminated in the revolution of 1830 which proved difficult for the Russian government to suppress. For the comments by Giers on this revolt see pp. 33–36, 40–41.

[6] General P. Kh. Vitgenstein was best known for his campaigns in the Napoleonic wars. After Kutuzov's death in 1813, he became for a while commander in chief of the Russian and Prussian armies. He also took part in the Russo-Turkish war of 1828.

[7] Alexei Andreevich Arakcheev was one of the most influential Russian ministers after 1818 when Alexander I abandoned his former liberal inclinations. War minister after 1806, Arakcheev introduced the hated system of military colonies.

II

[1] Friedrich Humboldt, German naturalist, like Litke, traveled widely and recorded his observations. He was chiefly interested in physical geography and meteorology.

[2] George Cuvier, French naturalist, specialized in zoology and paleon-

tology. He was also permanent secretary of the National Institute of France.

3 Nicholas himself went to the square and subdued the assembled mob with a dramatic speech: "What did you do yesterday? . . . You have shamed me before the whole world! . . . Come—if you wish to kill me with grief!" Constantin de Grunwald, *Tsar Nicholas I* (New York: Macmillan, 1955), pp. 121–122.

4 I. I. Panaev was later known chiefly for his critical and editorial work. In 1847 together with Nekrasov he revived *Sovremennik* (*The Contemporary*) which became an influential literary journal.

5 The Russian word for "to heat" or "to put fuel in" is *topit'*. Attse should have asked "topili?" meaning "Did you build a fire?"

III

1 For the experiences of Pushkin, Korf, Del'vig, and Kiukhel'becker in the Lyceum, see Ernest J. Simmons, *Pushkin* (Cambridge: Harvard University Press, 1937), chaps. 3 and 4.

2 The grand duke intended a pun. In Russian the word *dubina* means a club, but colloquially also a "blockhead."

3 M. B. Petrashevskii was the head of a circle which advocated ideas similar to those of Fourrier. In 1849 one of the meetings was raided, and the participants, among whom was Dostoievskii, were tried and eventually exiled to Siberia.

4 G. R. Derzhavin, a poet and dramatist, was prominent during the reign of Catherine the Great.

5 P. D. Kiselev as governor general of the Danubian Principalities supervised the administrative reorganization of the government. See pp. 123, 146.

6 G. A. Potemkin was a favorite of Catherine the Great and one of her ablest ministers.

7 V. F. Adlerberg was the closest friend and confidant of Nicholas I, who wrote of him in his will, "I love him like a brother . . . and I hope to keep him until the end of my days." Constantin de Grunwald, *Tsar Nicholas I* (New York: Macmillan, 1955), p. 138.

8 An English translation by Robert Hillyer can be found in Samuel H. Cross and Ernest J. Simmons, eds., *Centennial Essays for Pushkin* (Cambridge: Harvard University Press, 1937), pp. 225–226.

9 V. Zhukovskii was a prominent Russian poet, also known for his translations of western European romantic and preromantic poetry.

10 N. N. Novosil'tsev was a close companion of Alexander I and a member of the early reform circle.

IV

1 A. S. Griboedov, playwright and author, was sent to Persia as minister and was killed during a riot in Teheran in 1829.

[2] Albazin, a Russian fort on the Amur River, was surrendered to China in the Treaty of Nerchinsk in 1689.

EDITORS' INTRODUCTION TO PART TWO

[1] In the words of a member of a great Phanariote family: "*L'emporter sur ses compétiteurs et pour y parvenir, employer les moyens insideux dont l'usage n'était que trop encouragé par les Turcs, telle était la constante occupation des Grecs du Phanar: lutte incessante dont l'enjeu était toujours leur fortune et bien souvent leur vie.*" Nicolas Soutzo, *Memoires du Prince Nicolas Soutzo, Grand Logothète de Moldavie, 1798–1871* (Vienna: Gerold, 1899), p. 4.

[2] William Macmichael, *Journey from Moscow to Constantinople in the Years 1817, 1818* (London: John Murray, 1819), p. 107.

VI

[1] V. I. Tumanskii, in addition to his duties in the consulate, was also a poet. He published much, but was of only mediocre ability.

[2] The Skoptsy ("The Castrated") was a Russian religious sect founded in the middle of the eighteenth century by K. Selivanov who taught that "the only way to avoid temptation was to make it impossible for people to sin." Paul Miliukov, *Outlines of Russian Culture* (Philadelphia: University of Pennsylvania Press, 1942), III, 92.

[3] Dimitrie Cantemir joined Peter the Great in his unsuccessful campaign against the Turks and was forced to emigrate to Russia. He was the author of numerous historical works including *History of the Growth and Decay of the Othman Empire* (London: J. J. and P. Knapton, 1734).

[4] A. V. Suvorov was one of the greatest Russian generals. His brilliant victories over the Turks were the main strength of the eastern policy of Catherine the Great. In 1799 he was in charge of the army sent to Italy against Napoleon.

VII

[1] A desiatin is 2.7 acres.

[2] Giers here refers to the school of painting of the medieval period associated with the city of Suzdal'. Professor David Talbot Rice describes this school in the following manner: "Delicacy is perhaps the chief characteristic of the school, combined with reserve and good taste, both of which dominate throughout. But in later manifestations of Suzdalian work there is a tendency to weakness, which would be relieved by a more primitive, less refined touch. Suzdal', in fact, holds something of the same position in Rus-

sian painting as Siena holds in Italy. At their height both schools produced works of a delicate beauty which has seldom been surpassed, but their very excellence led rapidly to sterility which prevented further development or expansion." *Russian Icons* (London and New York: The King Penguin Books, 1947), p. 27.

3 For a list of J. F. Neigebauer's works on the principalities, see N. Iorga, ed., *Documente privitore la Istoria Românilor*, X (Bucharest: 1897), pp. lxiv, lxv.

4 Since the memoirs were never completed, this information is not available.

Index